Evaluating Theory-Practice and Urban-Rural Interplay in Planning

The GeoJournal Library

Volume 37

The titles published in this series are listed at the end of this volume.

Evaluating Theory-Practice and Urban-Rural Interplay in Planning

edited by

DINO BORRI

Department of Architecture and Urban Planning,
Polytechnic University of Bari, Italy

ABDUL KHAKEE

Department of Political Science,
University of Umeå, Sweden

and

COSIMO LACIRIGNOLA

Mediterranean Agronomic Institute,
Valenzano, Italy

SPRINGER-SCIENCE+BUSINESS MEDIA, B.V.

Library of Congress Cataloging-in-Publication Data

Evaluating theory-practice and urban-rural interplay in planning / edited by Dino Borri and Abdul Khakee and Cosimo Lacirignola.
p. cm. -- (GeoJournal library ; v. 37)
Selected papers presented at a workshop held at the Centre international de hautes études agronomiques mediterranéenes (CIHEAM), Valenzano (Bari), Nov. 1993.
ISBN 978-94-010-6297-8 ISBN 978-94-011-5462-8 (eBook)
DOI 10.1007/978-94-011-5462-8
1. Regional planning--Environmental aspects--Congresses. 2. City planning--Environmental aspects--Congresses. 3. Environmental policy--Congresses. 4. Sustainable development--Congresses.
I. Borri, D. (Dino) II. Khakee, Abdul. III. Lacirignola, Cosimo.
IV. Series.
HT391.E93 1997
307.1'2--dc21 96-39537

Printed on acid-free paper

To the memory of Professor Giovanni Grittani

CONTENTS

Preface

This volume contains a selection of papers presented at the second workshop on Evaluation and Planning held at Centre International de Hautes Études Agronomiques Méditerranéennes (CIHEAM) in Valenzano (Bari) in November 1993. The workshop was financially and otherwise supported by the School of Engineering, Bari Polytechnic; the School of Agriculture, University of Bari; and CIHEAM.
The publication of this book was made possible by to the efforts of the contributing authors. Several other persons have provided invaluable support for the workshop or the preparation of this volume. One of these is Patsy Healey for her fascinating challenge to Andreas Faludi's most recent arguments about rational planning theory. Another is John Friedmann whose lecture at the workshop presented world future scenarios depicting interaction between economic growth, social justice and ecological balance. Angela Barbanente provided marvelous support in organizing the workshop and editorial advice in the preparation of this volume. Jeremy Franks carefully improved the English and the clarity of all the papers. Carmelo Torre made a final editing of texts and images. We owe thanks to Maurizio Raeli for providing all the support services during the workshop and Claudia Baublys for her excellent help with various administrative issues with regard to the workshop and publication of this book.

This book is dedicated to the memory of Professor Giovanni Grittani, Professor of Land Economics, University of Bari. He was one of the organizers of the workshop on Evaluation and Planning and would have been one of the editors of this book. His sudden death in May 1995 robbed us of his intellectual and professional contributions and the work on this book has been profoundly affected by his absence.
Professor Giovanni Grittani (1943 - 1995) began his university career in 1971 as a lecturer in Rural Evaluation and Accountancy; and in Agricultural Economics and Policy at the University of Bari. He was promoted to professor of Rural Evaluation and Accountancy in 1986. From 1990 he was the Director of the Evaluation and Rural Planning Department at the Bari Faculty of Agriculture. At the same time, Professor Grittani was a member of Centro Studi di Estimo e Economia Territoriale and of the editorial boards of Genio Rurale, Politica Agraria and Medit.
Professor Grittani's prodigious scientific production deals mainly with agricultural economics and rural evaluation. In recent years his research came to focus on issues concerning the evaluation of public environmental goods. His recent publications include a handbook Estimo: Teoria, Procedure e Casi Applicativi (1994, written together with Professor M. Grillenzoni), and Estimo ed Esercizio Professionale (1995). Professor Grittani was particularly involved in the promotion of the professional status of graduates of Agricultural Science specializing in rural and regional planning.

About the editors

Dino Borri is a civil engineer and a Professor of Urban Planning Techniques at Bari Polytechnic. He is currently the head of the post-graduate School of Urban and Regional Planning at the University of Bari and the President of Fondazione Astengo, Istituto Nazionale di Urbanistica, Rome. He has written several books and articles on the application of Artificial Intelligence to spatial planning.

Abdul Khakee has a PhD in Economics as well as Geography and is currently a Professor of Urban Planning at the Department of Political Science at Umeå University. He has been a visiting researcher at the University of Delaware, the University of Melbourne, Bari Polytechnic and the University of Lisbon. He is the author of several books and many journal articles on development planning, planning theory, local government and cultural economics.

Cosimo Lacirignola has a degree in Agriculture Science and is currently the Director of Centre International de Hautes Études Agronomiques Mediterranéennes, Valenzano (Bari). He is also the Director of Medit, an international review of Economics, Agriculture and Environment. Between 1982-83 he was responsible for the European Commission's Programmi Integrati Mediterranei, and between 1983-87 he was principal administrator of the General Secretariat of CIHEAM in Paris.

INTRODUCTION

Dino Borri
Abdul Khakee
Cosimo Lacirignola

This book explores the evaluation of theory-practice and growth-conservation interactions in planning. These issues are becoming more important because evaluation methods have paid too little attention to decision-making aspects of uncertainty, value conflicts, ecological imperative, use of non-expert knowledge and citizen participation. Too few studies show the results of evaluation, and as a result planners and decision makers do not make use of the evaluation results as intended (Ormala, 1987). Too often, evaluations are used to justify expert opinions (Weiss, 1981). Further soft theories and soft methodologies influence evaluation in theory and practice as well as of growth-conservation interactions. The latter involves a conglomeration of pluralistic values. Relatively few attempts have been made to understand the value conflicts. When such attempts have been made, growth perspectives have often prevailed (Hajer, 1992).
Two set of changes have far-reaching implications for theory-practice and growth-conservation interactions: (1) the conception and the role of evaluation, and (2) the methodological framework.

Changing conception and role of evaluation

Following the financial pressure on the public sector, different reforms have increased the effectiveness of public agencies. In several cases, production responsibilities have been transferred to the private sector. Public sector reforms include the introduction of management ideas and market like arrangements including management by goals, separation of orderer and performer functions and contracting out the production to private companies. In outright privatization, production and distribution of collective services are managed by private agencies. However, the public-good nature of these services requires that government exercises control to ensure social justice in distribution and guarantee that explicit attention is paid to externalities. Several countries have replaced direct control through statutory requirements and public standards by indirect control through discretionary laws (Houlihan, 1992; Elander and Montin, 1990).

Evaluation plays a vital role both in achieving efficiency and ensuring that production and distribution take place according to public goals.

Since the late 1980s the environmental imperative has produced a radically new approach for development strategies. The emphasis on sustainability, increasing realization of linkages between global and local environmental changes and material constraints on economic growth no longer meant that only government ensured the preservation of natural resources and environmental quality (Newby, 1990). Davoudi argues here that ensuring sustainable development and ecological balance requires the simultaneous involvement of all actors - government, the market and citizens. This requires a different conception of evaluation which considers the objectives of a range of stakeholders who are not only interested in the outcome of a program or policy but also in its implementation in which they are important actors (Minnery et al, 1993).

In recent years new approaches to urban and regional planning are evolving, linked both to changes in environmental thinking as well as to territorial perspectives which emphasize the interplay between urban and rural elements. Evaluation studies involving urban and rural development have often been carried out from the urban growth perspective (Lockeretz, 1991; Grittani, 1989). Bearing in mind development in the industrialized and non-industrialized worlds, we must examine urban-rural development from the rural point of view too. This requires significant changes in the ways we make judgements and establish causality between urban and rural elements.

Citizen participation is an underlying concern in the communicative perspectives of planning research and an important aspect of developing sustainable development discourses (Friedmann, 1987). Although 'citizens' represent a starting-point in the parliamentary power relations at the national, regional and local level, citizen perspectives have been almost absent in the international evaluation discourse. Citizens need evaluation to decide whether politicians really carry out their duties in the public interest. In a representative democracy citizens are the clients and politicians their representatives vis-à-vis bureaucracy, business and other organised interests in the society. Information about political decisions and their impacts are vital for citizens to carry out their electoral duties (Vedung, 1991). The recent debate about citizen involvement in the production of collective services and the need to change consumer behaviour and moral responsibility to achieve environmental sustainability extends citizens' role. Evaluation of these issues involves fundamental changes in the epistemological premises and in the dispersion and effective use of information among the widest possible range of stakeholders.

Changing methodological framework of evaluation

A common thread running through many definitions of 'evaluation' is the comparison of performance and objectives, notwithstanding suggestions about "goal-free-evaluation" (Scriven, 1973). Evaluation involves a systematic examination of the effects of either proposed or ongoing or implemented policies and programmes. Evaluation methods are thus divided into ex-ante, continuous and ex-post evaluation. The planning literature has often focused on ex-ante evaluation. With the increasing emphasis on implementation and monitoring phases of planning, new tools for continuous and ex-post evaluation have been developed. Some researchers make a further distinction between discrete and continuous methods. The former assume that alternatives are explicitly given whereas the latter only assume the existence of goals and constraints but not the alternatives (Nijkamp, Rietveld and Voogd, 1990). Ex-ante evaluation methods include "monetary evaluation methods" (e.g. cost-effectiveness analysis, cost-benefit analysis), "overview methods" (e.g. planning balance sheet, community impact analysis) and "multi-criteria methods" (e.g. goal-achievement analysis, environmental impact analysis) (Faludi and Voogd, 1985). Continuous and ex-post methods, assessing the actual impact of policies and programmes, examine actual and intended results and system models which may focus on structural changes, achievement, feed-back, and so on (Vedung, 1991). In the case of the evaluation of environmental resources, there has been an increasing use of the so-called contingent valuation methods which attempt to measure individual and social preferences in terms of monetary units. These methods differ from those which make a more direct and substantial appreciation of environmental assets (Mitchell and Carson, 1989).

Though the academic discourse of ex-ante evaluation has been concerned mainly with technical aspects, increasing attention is being paid to conflicting objectives, accountability to diverse interests, and the identification and assessment of intangibles (Voogd, 1983). Ex-post models have often put emphasis on rational behavioural aspects, top-down control and relationships between politics and bureaucracy. More recent developments pay increased attention to side-effects, the distinction between various categories of clients and the multi-disciplinarity of interests (Vedung, 1991). Thus we see a closer interplay and a longer run convergence between ex-ante and ex-post evaluation techniques.

So evaluation is not only an intellectual and instrumental but a political and institutional process. This acknowledgement requires us to study not only issues of efficiency versus social justice but also questions which fundamentally change the framework for evaluation (Pearce, 1993). How is efficiency concerned with community values and issues of need? Whose goals and values are relevant? Should we consider unofficial goals and look at side effects? How can we distinguish and choose between the many sources of public goals? How can we account for the change in goals over time? How can we make explicit the trade-offs between various goals? These issues require evaluators to assess the arguments invoked in favour of various decisions. Evaluation allows negotiations between multiple parties, which Voogd (below) has called the "will-shaping" process, part of a more general analysis of the discourses and discourse-making in planning (Healey, 1993).

Third, the methodological framework of evaluation is changing by extending the object of evaluation from projects, policies and operational plans to strategic plans and plan-making processes. Accordingly, Faludi and Altes here propose the term 'performance' as opposed to 'conformance' and imply thereby that departures from plans do not necessarily indicate failures. Similarly Lichfield suggests that evaluations of strategic plans are general, indicative, and involve considerable qualitative judgement.

Planning-process evaluation leads us to investigate how values in deliberative settings arise, how relationships, responsibilities, networks and collective memory can be transformed and how conflicts and disputes may be resolved (Forester, 1992). These issues call for more than the customary techniques. Moreover planning processes involve far more than the stated intentions (for example, of planning legislation). In short, evaluation of strategic plans and planning processes involve pluralistic investigations of plan contents and planning processes - both seen as arenas for struggle in which different interests compete (Khakee, 1994).

Implications

So what do these developments in our understanding of evaluation and the methodological frameworks of the theory-practice and growth-conservation interaction imply?

Consider four implications concerning the theory and practice of evaluation:

1. Methods for evaluation need to be more transparent than ever before. For stakeholders to be directly involved in policy actions (especially in bringing about environmental sustainability), evaluation methods must be accessible and easily manageable. Evaluators have to pay more attention to the ways of communicating with decision-makers as well as the public at large, not only to the results but also to the procedures of their analyses.
2. As indirect control by government increases, evaluation becomes all the more crucial. The three major causes of insufficient use of evaluation need to be remedied. First, the production-related shortcomings concern the lack of relevance and poor quality of evaluation results. Second, the potential users of evaluation results, namely decision-makers and administrators, are often not result-minded and do not appreciate the value of evaluation in decision-making. Third, imperfect communication between producers and users of evaluation studies can be remedied by punctuality, accessibility and proper channels for disseminating the results of evaluations.
3. We must clarify major concepts and conflicts between different interpretations of these concepts in the face of the increasing use of evaluation in environmental policy. The notion of 'environmental quality' as a yardstick to measure development has increasingly gained in importance (Voogd, 1994). The integration of ecological aspects in socio-economic planning faces substantial limitations because 'ecology' can be interpreted in different ways - as a geographical component of a physical system, a specific level of life-organization and so on. Similarly 'sustainable development' has been interpreted in several ways.
4. A fundamental issue in evaluation involves the values which motivate the identification of goals and which enable judgement. In the rational model, values were 'given' as 'exogenous' to the policy-making process. In our 'post-rational' thinking, values are derived endogenously in an interactive policy process. Evaluation becomes all the more difficult as the communicative process involves not only policy outcomes but also mutual learning and behavioural transformation. The shift towards the communicative approach to planning raises the need for dealing with 'real' choices as opposed to the implementation of 'abstract' statements in the traditional planning. When assessing individual values for social and natural goods, there is a need to go beyond the neo-classical social choice theory with its linear addition of individual behaviour to an action context which is contradictory, plural and non-transitive and which requires evaluation of measurable and unmeasurable assets (Sagoff, 1994)

Consider four implications of these issues for the growth-conservation interaction.

1. We must examine both the conflict and the complementarity between growth and conservation perspectives. Neoclassical economics implies that the loss of natural resources can be offset by man-made resources developed through technological advances. Some ecologists reject this contention and argue that the imbalance in the ecological system can only be remedied by a fundamental reversal of the growth ideology. Co-evolution of the economic and ecological systems requires careful examination of the limits to growth, inter-generational rights to environmental resources, time horizons for balanced planning, etc.
2. The growth-conservation interaction becomes more crucial as urban and rural development occurs within a specified territory. As urban and rural elements are integrated in a territorial plan, the role and nature of planning changes. Current territorial planning often has an urban growth perspective. Rural issues are often given a secondary status. If a more balanced urban-rural approach is to be developed, we must evaluate (a) citizen preferences regarding 'existence value' of agriculture, forests and other green areas, (b) the preferences of farmers and other occupational groups employed in the rural sector, (c) the real estate value and the impact of legislation and plans on this value, and (d) recreational values represented by the rural environment. These considerations are especially important in the light of the varied patterns of urbanization, cultural context and urban size. Evaluators must attend to both macro- and micro-perspectives of the growth-conservation interaction.
3. The conceptual differences and the status of conservation in relation to material growth issues require us to develop economic and ecological indicators of practical value. Evaluation studies combining growth and conservation concerns often pay lip service to the latter without properly investigating the true costs of neglecting ecological balance. Conventional performance indicators for sustainability pay little attention to locational conflicts, time horizons, and the consistency of multi-scale spatial applications (Burnell and Galster, 1992; Herzog and Schlottmann, 1993).
4. A constant struggle persists in the fringe areas of urban settlements and hence there is a special need to look at growth-conservation interaction. Infrastructure in the form of highways, oil pipelines, sewers, and electric lines influence fringe areas more significantly than evaluation studies often appreciate. Specific evaluation techniques for these areas are therefore needed.

The contents of this volume

This volume presents the papers from the workshop on 'Evaluating Theory-Practice & Urban-Rural Interplay in Planning' held at the International Center for Advanced Mediterranean Agronomic Studies, Valenzano (Bari), 18-20 November, 1993.
Theory and methods in Part One precede studies on evaluation in planning practice in Part Two, and case studies discussion on urban-rural interplay in Part Three. However, the distinction is far from rigid. Several methodological papers refer to case studies and many case studies discuss theoretical and methodological issues relating to the theory-practice and growth-conservation interactions.
The papers in Part One by the Dutch participants present specific methodological proposals for evaluation. Faludi and Korthals Altes distinguish technocratic and

sociocratic approaches to planning. In the former, authorities safeguard the public interest, whereas the latter pays greater attention to the views of others. They propose two models of interaction between the plan-maker(s) and subsequent decision-takers: a mechanistic one way flow of information and an interactive two way communication model. Voogd suggests that evaluation is not only an intellectual but also a political, social and institutional process. Evaluation must be accompanied by 'will-shaping' processes to familiarize people with alternatives and evaluation outcomes. Lichfield's paper addresses environment impact assessment or community impact analysis as applied to plans rather than to specific projects. Strategic Environment Assessment can be integrated with Strategic Plan Evaluation and Lichfield traces the institutional and methodological changes required.

The next three papers by the Italian scholars reflect the current interest of Italian evaluation researchers in the theoretical and practical limits of evaluation methods. Fusco Girard presents four major issues about multicriteria approaches in evaluating environmental policies. First, the present reorientation of the economy towards market management hinders sound environmental policy. Externalities reflect the market's inability to exploit resources in a sustainable way. Second, evaluation as so far applied has not succeeded in averting irreversible damage to the environment. Third, the economic value of environmental resources differ for developers, consumers, the community as a collective, and political interests. Fourth, evaluation has not yet been able to expose or even help to reduce conflicts among different interests, goals and values. Fusco Girard contends that multicriteria evaluation methods, focusing on multiple groups and diverse goals etc., are useful in mediating between different institutions and interests. Macchi and Scandurra discuss three 'paradigms': economic, ecological and planning. The blind trust in technological progress among neo-classical economists, contradictions and shortcomings in ecological information, and the rare success of urban and regional planning despite the technology at its disposal are the major sources of these limitations. Ingegnoli outlines the evolution of Italian environmental planning from its emphasis on landscape planning to its current concern for sustainable development. The author assesses present shortcomings involving conceptual contradictions regarding the concept 'ecology' and the relative immaturity of ecological science regarding its integration of physical and human systems.

The papers in Part Two emphasize the role of practice in evaluation. Forester's paper poses challenges for the evaluation of the deliberative and participatory plan-making process in which participants learn about values, and transform themselves with regards to relationships, alliances, networks, etc. Values are endogenous to the planning process, not external to it. So what values should be considered in the evaluation process, and how to delineate them? How to evaluate the results of planning when the output involves in part appreciation, understanding, mutual learning, and so on? Forester's paper raises more questions than it answers. Khakee's paper moves from the evaluation of strategic plans to that of strategic planning processes. He proposes a methodology for evaluating the process and presents its application to structure planning in Swedish municipalities. Roscelli refers to our architectural heritage which he views as scarce, unique, non-reproducible and not 'usable' in an ordinary sense. He indicates the gap between theory and practice and the epistemological problems of evaluating cultural heritage.

Barbanente and her co-authors discuss the difficulties of integrated transportation and land use planning especially as they involve environmental issues like sustainability. Besides conceptual problems, such planning involves conflicting goals of social justice, nature's rights and values and qualitative development. One response to these problems is to develop models for effective knowledge bases to enable proper interpretation of conflicting goals. The authors propose a model using 'expert knowledge'. The paper by Caiulo and his colleagues reflects upon the challenges of deliberative and participatory policy processes posed earlier by Forester. The authors explore two case studies: the north-south arterial road in Bari and the Pittachi road in Brindisi. The process leading to the realization of these projects left much to be desired. The absence of communication and interpretation of knowledge, the lack of fit with the rest of the infrastructure system, the legislative hindrance of public participation and the deliberate fragmentation of projects into smaller sub-projects were only some of the dismal aspects of this process. No wonder that the authors feel that current planning practice too often only pays lip service to environmental impact assessment and citizen involvement in achieving sustainable development.

The six papers in Part Three focus on environmental protection and urban-rural interplay in evaluation. Davoudi traces the history of environmental planning in Great Britain and illustrates the absence of a policy for long-term ecological balance with the example of the mining of whinstone in northern England. Sustainable development implies a new approach to the mediation of interests, new criteria for development, new ways of thinking about human responsibility, and significant institutional rearrangements. The specific focus should be on 'biospheric' quality, capacities and thresholds that limit the exploitation and destruction of resources and the ecological system. Also the paper by Healey and Shaw focuses on the role of British environmental policies in sustainable development strategies, with special attention to the land use planning system. The paper gives a contribution to the problem of operationalizing environmental consideration in the planning system, in the light of the meaning of environmental sustainability, the changes in economic and social life, and the evolution of new approaches to urban and regional policies.

The next paper by Franceschetti and Tempesta explores two neglected issues: the comparison of regional plan goals for landscape protection with people's preferences, and the impact of landscape protection laws on the value of real estate within national parks. De Fano and Grittani focus on the functions of 'peri-urban' areas where recreational, aesthetic, agricultural, semi-urban development interests interact. Planners often have a bias for growth and for creating empty spaces. De Fano and Grittani advocate a multicriteria approach employing multiple indicators for various functions of the peri-urban area. The following paper by Carbonara proposes a multicriteria model for evaluating agricultural and environmental resources of land. Carbonara's comprehensive model includes an analysis of growth resources (economic, infrastructure and settlement) and conservation resources (agricultural, recreational and ecological) and enables an analysis of maximum and minimum alternatives for different growth-conservation proposals. The last paper by Trisorio-Liuzzi and her colleagues takes up an often neglected topic in evaluation literature, the impact of natural hazards on urban and regional planning. How can urban and regional planning integrate risks associated with natural disasters? What evaluation methods are available for this purpose? The authors propose the use of thematic maps as a method for assessing risk conditions generated by natural phenomena.

References

Burnell, J.D. and Galster, G. (1992) "Quality of life measurement and urban size: An empirical note", Urban Studies 29, 727-36.

Elander, I. and Montin, S. (1990) "Decentralization and control: central-local government relations in Sweden", Policy and Politics 18, 165-180.

Faludi, A. and Voogd, H. (eds.) (1985) Evaluation of Complex Policy Problems, Delftsche Uitgevers Maatschappij, Delft.

Forester, J. (1992) "Envisioning the politics of public sector dispute resolution", Studies in Law Politics and Society 12, 83-122.

Friedmann, J. (1987) Planning in the Public Domain: From Knowledge to Action, Princeton University Press, Princeton.

Grittani, G. (1989) "La valutazione monetaria del territorio rurale nei processi di pianificazione urbana e regionale", in Barbanente, A. (ed.) Metodi di Valutazione nella Pianificazione Urbana e Territoriale. Teoria e Casi di Studio, Consiglio Nazionale delle Ricerche, Bari.

Hajer, M. (1992) "The politics of environmental performance review: Choices in design". In Lykke, E. (ed.) Environmental Performance Review, Belhaven, London.

Healey, P. (1993) "The communicative work of development plans", in Khakee, A. and Eckerberg, K. (eds.) Process & Policy Evaluation in Structure Planning, Swedish Council for Building Research, Stockholm.

Herzog, H. W. and Schlottmann, A.M. (1993) "Valuing amenities and disamenities of urban scales. Can bigger be better?", Journal of Regional Science 2, 145-65.

Houlihan, B. (ed.) (1992) The Challenge of Public Works Management, International Institute of Administrative Science, Brussels.

Khakee, A. (1994) "A methodology for assessing structure planning process", Environment and Planning B 21, 441-451.

Lockeretz, W. (1991) "Secondary effects on Midwestern agriculture of metropolitan development and decrease in farmland", Land Economics 65, 205-216.

Minnery, J. et al. (1993) "Evaluation in urban planning", Australian Planner 31, 8-13.

Mitchell, R.C. and Carson, R.T. (1989) Using Surveys to Value Public Goods - The Contingent Valutation Method, Resources for the Future Inc., Washington D.C.

Newby, H. (1990) "Ecology, amenity and society", Town Planning Review 61, 3-13.

Nijkamp, P. Rietveld, P. and Voogd, H. (1990) Multicriteria Evaluation in Physical Planning, North Holland, Amsterdam.

Ormala, E. (ed.) (1987) Evaluation of Technical Research and Development : Experience of Practices and Methods in the Nordic Countries, Nordic Cooperative Organization for Applied Research, Helsinki.

Pearce, B. (1993) "Evaluating land-use planning: the importance of impact research", in Khakee, A. and Eckerberg, K.(eds.) cit.

Sagoff, M. (1994) "Should preferences count?", Land Economics 70, 127-44.

Scriven, M. (1973) "Goal-free evaluation", in House, E.R. (ed.) School Evaluation: The Politics and Process, McCutchan, Berkeley.

Vedung, E. (1991) Utvärdering i politik och förvaltning (Evaluation in Politics and Administration), Studentlitteratur, Lund.

Voogd, H. (1983) Multicriteria Evaluation for Urban and Regional Planning, Pion, London.

Voogd, H. (ed.) (1994) Issues in Environmental Planning, Pion, London.

Weiss, C.H. (1981) "Measuring the use of evaluation", in Ciarlo, J.A. (ed.) Utilizing Evaluation: Concepts and Measurement Techniques, Sage, London.

PART I

THEORY AND METHODS

1 EVALUATING COMMUNICATIVE PLANNING

Andreas Faludi[1]
Willem Korthals Altes

1.1 Introduction

Studying plan implementation seems simple. Either plans work, or they don't! However, consider a plan that has been implemented, but does not have the desired effect. Is it a failure? Hardly! After all, the plan has been implemented. In fact, the shoe pinches somewhere else: the plan has been flawed. As Bardach (1980, 141) says, we should not "...necessarily count as a failure of the implementation process a result that originates in a more fundamental conceptual defect in the policy design. If a policy of rent control, for instance, leads in the long run to housing shortages and deterioration of the housing stock, it would probably be mistaken to blame those outcomes on the way the policy was implemented. ... 'Good' implementation cannot by itself offset the ill effect of 'bad' policy any more than a more perfect compass and straight edge can help us to square the circle." To the extent that blame needs to be apportioned, it is to the planners for their failure "...to recognize their conceptual mistakes and to take constructive measures to correct them. It is necessary to emphasize this special case of failure-as-success because policy making is often not a straightforward matter. There is inevitably a certain amount of trial and error, and sometimes the opportunity for trial and error is -- or ought to be -- built in. A successful implementation process, therefore, not only avoids known pitfalls but seeks better and perhaps unpredictable paths to new and perhaps unforeseen destinations."

The upshot is that non-conformance of outcomes with the plan, conventionally counted a 'failure', is not necessarily a bad thing. Also, implementation of a plan in conformance with original intentions, normally regarded a 'success', can lead to undesirable outcomes. To avoid them, what we would wish is what Bardach suggests: whilst being applied, plans should be reconsidered.

This is far from revolutionary. However, what, then, is the criterion according to which we should evaluate the quality of plans? Apparently, the means-ends scheme underlying mainstream evaluation, in which conformance between plan and final outcomes is the test of effectiveness, does not always apply. The reason why this scheme is often

D. Borri et al. (eds.), Evaluating Theory-Practice and Urban-Rural Interplay in Planning, 3–22.

recommended seems that most evaluators think about projects being implemented. However, strategic plans are different. Such plans are frameworks for action and need to be analysed for their performance in helping with subsequent decisions.

This is not new either. There is a small but growing literature on 'conformance' versus 'performance' (Fudge, Barrett 1981; Mastop 1987; Alexander, Faludi 1989; Faludi 1989a; Mastop, Faludi 1993). Here we add a new element to this. We look at the interaction between the maker of a plan and the subsequent decision makers as a process of communication, albeit one using texts rather than face-to-face contacts as the medium of exchange (see also Korthals Altes 1993).

Talking about 'communicative planning' is not novel (Friedmann 1973; Forester 1989; Zonneveld 1991; Edwards 1991; Healey 1993). Mostly, the focus is on planning as involving face-to-face communication. There is also attention for the rhetorical quality of plans. Our focus is different. We concern ourselves with how to evaluate the products of communicative planning. Specifically, we want to find out whether these products help with making sense of decision situations, more particularly with seeing them against the backcloth of the wider field of choice.

First, we go briefly through the argument concerning performance. It involves a shift in perspective that eliminates the quandary which Bardach describes. This is what section 1 below is about. Section 2 shows that the 'quality of decisions' cannot be established other than by looking at the arguments adduced in their favour. This will lead us into discussing communication in Section 3. There are two models of communication. They presuppose different patterns of relations between plans and subsequent decisions. One pattern is hierarchical: the 'context of command' in which project planning is appropriate. The other situation is characterised by symmetry: the 'context of accommodation' in which strategic planning is more suitable. This leads finally, in Section 4, to the distinction between two types of evaluation: conventional, and our type that focuses on the performance of plans during subsequent negotiations. The article ends by discussing various approaches to establishing plan performance.

1.2 The purpose of planning

Planning thought often evolves around pairs of concepts. Undoubtedly, blueprint planning and the process approach is the most popular pair (Faludi 1984, first published 1973, chapter 7). Foley (1963; 1964) uses the somewhat different concepts 'unitary' and 'adaptive' approach. Claiming that the 'I.O.R. School' (Institute for Operational Research) represents a radical departure from previous thinking about the purpose of planning, Faludi and Mastop (1982) use a similar pair of concepts. According to previous thinking, planning is the preparation of plans by experts. Representing a broad view and a bright future, such plans need to be followed. This assumes that the plan caters to true needs, hence the efforts going into surveys and forecasts. According to the I.O.R. itself, planning "...is concerned not so much with producing a plan as with gaining a better understanding of the problems with which we are faced now and in the future, in order that we can make better decisions now." (Centre for Environmental Studies 1970, 15-16)

There is a real difference here, compared to conventional views, including the classic rational planning model as epitomised by the systems approach (Faludi 1987, 91-92).

According to the new view planning should never set itself up as an alternative to ordinary decision-making, but should help in making ongoing choices. At the same time, planning raises choice to a higher level of awareness. Indeed, this is what distinguishes planning from non-planning. As Friend and Jessop (1977, 110) write: "...any process of choice will become a process of planning [or: strategic choice] if the selection of current action is made only after the formulation and comparison of possible solutions over a wider field of decision relating to certain anticipated as well as current situations."
Drawing inspiration from the I.O.R. school, Van der Valk (1989, 419) includes views of society and the state in distinguishing between a 'technocratic' and a 'sociocratic' approach to planning. The former assumes a strong role for authorities in safeguarding the public interest. As the experts, planners get considerable say. Everything evolves around the Plan. Those implementing it require no discretion or room for negotiation. The Plan has taken care of everything. The approach implies a plan-led system. The sociocratic approach, however, pays attention to views of others. Authorities are not the only ones called upon to act in the 'public interest' and not above other actors either. This leaves room for negotiations. The role of planners is less central than in the technocratic approach. Plans concern general lines of development. This view is more flexible, meaning the plan can be reconsidered.

Figure 1.1: Differences between Technocratic planning and Sociocratic planning

	Technocratic planning	Sociocratic planning
planning subject	monolithic	coalition
role experts	linchpin	one out of many
centralization decisions	great	small
plan as product	dominant	relative
form of plan	blueprint	indicative
measure of effectiveness	conformance	performance
scope	comprehensive	selective
notion of rationality	absolute	contextual
planning process	linear	cyclical

This resembles 'communicative' planning. Thus, Friedmann discusses what he calls 'transactive planning' (1973) in which the problem is "to improve the *quality of the action*" (1969, 311), a notion not unlike ours below of the 'quality of decisions'. Friedmann also takes a stand against technocracy: "The widespread notion that plans *ought* to get accepted and that, when they are not, the failure is one of communication, rests on the technocratic fallacy that planners' proposals are inherently superior to actions that result from the unaided decisions of non-planners." (1969, 311) Friedmann sees a good plan as "a drifting cloud" (1965, 39). It is not an inviolable document, like Holy Scripture, but a fleeting summary of current knowledge, expectations and goals.

Along similar lines, Forester (1987) states that actions fit in with arguments which on their part fit in with strategies. Arguments can be used in justifying decisions. So to establish the quality of the action, it is necessary to analyse the arguments invoked in favour of the underlying decision. Now, these works relate mainly to planning as a process. Two recent papers on communicative planning focus on the qualities of plans as such. The first is Healey (1993) on the 'communicative work' of British development plans. The focus is on how democratic values are reflected in such plans. Healey describes a plan as the product of interaction between a range of parties. The plan may become a point of reference for continuing interaction within which discourses may evolve. Each of these discourses has its own story line, and plan preparation thus involves 'making story lines'. More often, however, pre-existing 'stories' and strategies are consolidated and translated into reference criteria for future decisions. A plan is also "an arena of struggle".

Rather than taking the implications for the evaluation of the communicative work of development plans on board, Healey focuses on the existence of several discourses in each of the plans studied. She concludes that the communicative work which plans perform is to inform readers "who can understand" about land allocation and performance criteria. The plans as such are inserted into wider debates about development management, offering rhetorical strategies. About the plans studied, she then concludes that all of them perform argumentative work. However, none "...presents the plan within the context of an open debate among all those 'with a right to know'." (102)

Democracy, though, "...has to absorb the diversities and the differentiation which characterise our contemporary 'post-modern' culture. This has to mean recognition of the plurality of discourse communities. With respect to plan-making, it means explicitly addressing the challenge of 'interdiscursive communication'." (103)

"A democratic plan may thus be recognisable not so much by its aims, or its policies, or its 'distributive justice'. Rather its identifying character is likely to be its *tone*, expressing the experience of interdiscursive discussion." (103)

On similar lines, Throgmorton (1993) describes planning as a rhetorical, rather than an objective, technical activity. This solves the contradiction between the self-image of planners as objective technicians and their actual role which is highly political. "Rather than thinking of rhetoric as gloss or seduction, planners should regard it as the study and practice of persuasion, and recognize that persuasion is constitutive." (335)

This approach builds on three principles:

- Plans, analyses, and in fact the stories in plans are always addressed to someone, so the audience is important.
- Planning-related utterances are replies to other utterances, so we always argue in the awareness of differing or opposing views.
- The meaning of such an utterance is beyond the control of the speaker, so we must think about this "play of meaning" and about how audiences reconstruct meanings.

Throgmorton emphasises the role of planning tools as efforts to persuade specific audiences in specific contexts to accept proposed explanations, embrace inspiring visions, and undertake recommended actions. He acknowledges that such persuasive efforts take place in "...the context of a flow of utterances, replies, and counter-replies". Each of them is likely to be interpreted in diverse and often antagonistic ways.

"Audiences can assign different meanings to key terms, fill gaps in the original analysis, and choose to read either with or against the analysis." (335)

Rhetoric is constitutive for planning. The very way in which planners write and talk shapes their characters and their relations with their audiences: "Each time planners write or speak, they create ideal readers or listeners, who actual readers or listeners -- as objects of planning or even participants in it -- may or may not be, or may or may not choose to become." (336)

Throgmorton argues, then, that planners should strive to "...find a rhetoric that helps to create and sustain a public, democratic discourse. This should be a persuasive discourse that permits planners (and others) to talk coherently about contestable views of what is good, right, and feasible. Planners should strive to create arenas that facilitate and encourage just such a persuasive, public discourse." (336)

So, "...planners should surrender any further pretence to neutrality, objectivity, and universal truth... Surrendering the pretence ... should not lead them to the extreme of defining planning as just another form of politics gone amok. Instead, planners should embrace persuasive discourse and political conflict to realize that survey 'results' are, like all alleged planning facts, inherently tropal and contestable. They must be scientific *and* rhetorical, professional *and* political, because they, like all other planning tools, configure the planning arguments." (344)

This literature focuses on what planners do in their day-to-day work (Healey 1992), with an emphasis on how they can promote democratic values. Now, it may be apposite to remind ourselves that planners also make plans. With Forester (1987, 163) we describe planning as "...the *practical anticipation of potential project (or policy) implementation*". Plans are the products of such anticipation in the form of a text. It is certainly relevant to ask not only whether planning promotes democratic values, but also whether plans work or not. So far, though, the concern in this range of literature for plan evaluation has been limited (Krumholz, Forester 1990, 244-248). The focus is mostly on the evaluation of the quality of the actions and judgements of planners as such, and not on the evaluation of the success of plans and planning in placing decisions, as the decision-centred school would have it, in their wider field of choice (Friend, Jessop 1977). In this paper, we take note of the rich literature on communicative planning, but our focus is squarely on how to evaluate, to paraphrase Healey, the 'planning work' that it performs.

1.3 The quality of decisions

Given that the purpose of planning is not to impose plans on others but to help to improve the quality of subsequent decisions, planning needs to be evaluated in terms of whether it does precisely that. The question is then how we can measure the quality of decisions.

The answer seems deceptively simple: a decision is good if it conforms to whatever substantive criteria we care to invoke. At best, this is a tautology saying that a decision is good if it is good. At worst, like the saying 'My country right or wrong', it signals a self-regarding attitude. No, determining the quality of decisions requires us to look at how they relate to the situation at hand, whether they seem reasonable responses to the challenges which these situations hold, whether they are superior to any of the alternatives that present themselves. In a nutshell: the quality of decisions depends on the quality of the arguments adduced in their favour. We speak about these arguments as the justifications of decisions.

We can of course 'argue things out' with ourselves. However, the more usual situation will be that we justify decisions in the presence of, and for the benefit of, others.[2] Justification therefore presupposes the existence of an arena, with actors who for one reason or another claim to have a stake in that decision. They critically assess the arguments adduced, which is why they figure as the critics in our scheme.

This will take us in due course back to communication, but first we explore what this all means for evaluation. The implication is that the purpose of planning -- improving decisions -- can only be fulfilled by analysing the justifications of decisions. Evaluating planning, so conceived, therefore means establishing the extent to which planning improves the justifications of decisions.

There are various ways in which justifications might be improved. There are formal aspects, like consistency in the use of concepts, and the validity of inferences drawn. There are also criteria of relevance, for instance, the empirical referents in the argument leading to a decision must be valid. Following Habermas, Forester (1980, 277) suggests four criteria for assessing the 'speech acts' (like the justifications of decisions) of planners. Planners should speak comprehensibly, sincerely, legitimately and truthfully.

What follows from this list of criteria is that not all improvements to the justification of decisions qualify as being improvements due to planning considerations. Although one might wish that planning lived up to these requirements, consistency, relevance, truthfulness, sincerity and the like are not an exclusive domain of planning. For improvements in the justification of decisions to be due to planning considerations, they must relate to what Friend and Jessop have been quoted above as describing as the wider field of choice. We conclude: the purpose of planning is to improve the quality of decisions in terms of whether their justifications relate to the wider field of choice. Planning is simply a prior investment so that decision makers can take a broader view, that they are more aware of what it is they are doing.

This means that evaluation must home in on a specific part of the manifold pattern of interaction around the justification of decisions: the interaction between the plan maker and whoever is responsible for taking subsequent decisions.[3] More specifically, the evaluator must look at whether the plan maker has succeeded in conveying to the decision maker a better understanding of his or her wider field of choice. Note that the decision maker can be the same as the plan maker, but then in the capacity of an actor responsible for taking decisions.[4] In fact what happens is that the plan maker formulates a text, proposing that the subsequent decision maker (including him- or herself!) takes account of it. More precisely still: the plan maker provides an input into the justification of follow-on decisions.[5]

However, not only the decision maker may use the plan. Any actor in the arena within which justification takes place can do so. Now, it is well-nigh impossible for the plan maker to identify all such instances where the plan may be used beforehand. For instance, it is impossible to anticipate whether an action group or several might raise one or several issues or employ an argument derived from the plan, which adds to the uncertainty surrounding planning and, below, we argue that, as a group, users of plans are diffuse for still other reasons. First we return to the communication between plan-makers and those concerned with subsequent decisions.

1.4 Communication

We have argued that, where it draws on a plan, the justification of decisions involves communication between the plan maker and whoever is responsible for those decisions. There are various models for conceptualising communication. The difference is that, in the first model, communication can be meaningful only insofar the codes of sender and recipient conform, whereas according to the second model meaningful communication is possible even when the codes of authors and readers of texts differ.
So the first model conceives communication as a flow of information from a sender to a recipient (Witteveen 1992). The recipient is able to understand the message, because the sender and the addressee are using the same codes to construct and receive the message. The addressee gives feedback. This makes fine-tuning of codes possible. Be that as it may, if the information received is substantially different from what is intended, then it is said to be distorted. There is 'noise', and something needs to be done about it. The implication for planning is clear: if the plan does not get across as intended, then communication is distorted. (Figure 1.2)

Figure 1.2 (Source: Witteveen, 1992)

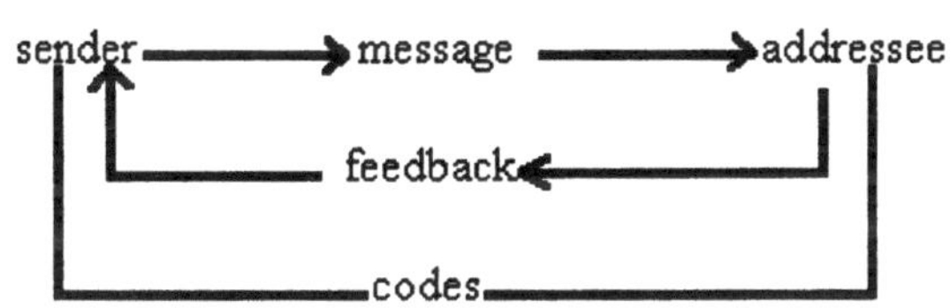

The second model of communication is less mechanistic. It sees communication as the double construction of text by both author and reader (Eco 1979, 5-11). Fundamental to this model is that communication can be meaningful, even where the codes of author and reader diverge from each other. The author of a text tries to imagine the frame of mind of the reader, and to tailor the message accordingly, thereby using his or her own code. The reader in turn forms an image of the author, interpreting the text in this light by using his or her own codes. When (as is often the case) the codes of the reader and author differ, then the meaning of the text constructed by the reader will necessarily differ from the text constructed by the author. (Figure 1.3)
The double construction of texts is a necessary part of communication and not a distortion. Fortunately, there are not only disadvantages to this which, according to the mechanistic model we would call information loss during transmission. The interpretation by the reader of a text can be superior to the one intended by its author. In particular, the reader will be better able to ascertain the significance of any text to his or

her situation. The latter is highly relevant to the specific form of communication that we are concerned with, that between the maker of a plan and subsequent decision makers. Decision makers are more aware of their situations than planners can hope to be when drawing up a plan.

Figure 1.3 (Based on Witteveen, 1992; see also Eco, 1979)

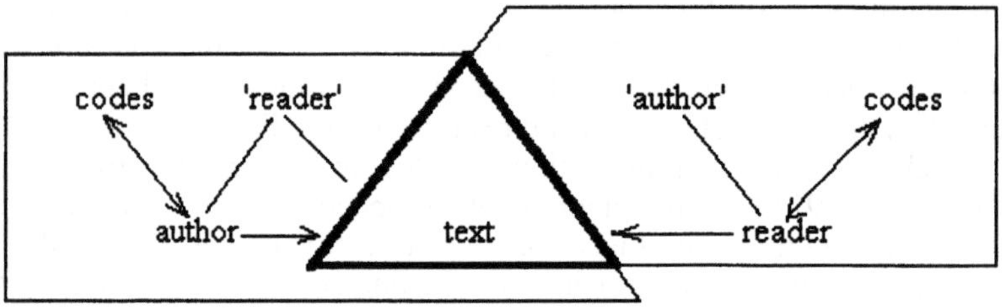

There are extreme cases, though, in which conceptualising writing and reading as two independent processes connected only by a text (Witteveen 1988) leads us nowhere. They are the cases where all codes of plan makers and plan users are different. Under these conditions, any odd text could substitute for the plan. The meaning that the users of a plan construct from it -- the 'plan-in-use' (see below) -- bears no relation to the meaning that the plan maker has attached to it. It follows that the codes of the plan maker and subsequent decision makers must have some resemblance to one another. The persistence of codes is a precondition, therefore, of increasing planning knowledge, but it is not a measure of performance. After all, persistence of codes does not contribute directly to improving the justification of decisions. The contribution is rather indirect in that a common code that persists over time enables us to learn from past experience.[6]

In psycholinguistics it is common to portray the role of the author in the interpretation of texts as trivial (Kreuz, Roberts 1993). However, the second model of communication has its roots in hermeneutics rather than psycholinguistics. More generally speaking it is rooted in a view of the study of society as a philosophical, rather than a 'scientific' enterprise based on causal explanation. One of the authors in this tradition, Winch (1973, first published 1958) states that the realm of human action can be understood only in terms of the concepts and conventions -- in other words, 'the codes' -- of the actors themselves. Following Wittgenstein's *Philosophical Investigations*, rule-following is not seen as an interpretation, but as a situated construction -- as a 'form of life' (see also Van Gunsteren 1976, 109-138).

However, in Winch's model, there is no role for an observer using his or her own codes and playing a different kind of 'language game' (Pitkin 1972, 250-272). Thus, for evaluation Winch's perspective has to be broadened, taking account of the evaluator as an observer. The evaluator has his or her own codes as to the 'purpose of planning', as we have shown above.

Although the second model is obviously more appropriate for evaluating communicative planning, the evaluator-as-observer may retort 'who cares?' Whatever the 'model of communication', a plan must be adhered to. After all, it represents a long-term view of what is in the 'public interest', and so it must have teeth! Clearly, this conjures up a pattern of relations between plans (or rather their makers) and subsequent decisions (or rather those responsible for them) that differ from what we envisage. We turn to discussing such patterns.

There are two opposites which, following Cronbach et al. (1980, 83-94), we describe as 'context of command' and 'context of accommodation'. The analogy with views of planning above is only too evident. In the context of command the technocratic view prevails. In the context of accommodation the sociocratic view does. Each implies its own form of planning and, as will transpire below, also of plan evaluation.

In a context of command that is suffused with a Platonic image of concentrated power and responsibility, officials are expected to make plans and execute them. Planning, then, is 'The Quest for Control' (Van Gunsteren 1976). Communication is mechanistic, as in the flow model above. Albeit implicitly, most evaluation takes place based on the assumption that a 'context of command' prevails. In terms of communication theory, there are some huge underlying assumptions. The most important assumption is that, in communicating, the makers and users of a plan use the same codes (Witteveen 1992). What a plan communicates are commands. Commands establish a direct relationship between addresser and addressee (Van Loon 1958). This relationship is a binary one: it can take the form of either conformity or disconformity between the command (plan) and the subsequent decision, or rather the act based on it. Disconformity always entails a loss of information. 'Noise' constitutes an obstacle to communication, so experts must make themselves crystal-clear. Evaluation in this context is about the implementation of the plan, and nothing else.

The relevant form is the project plan. It complements the technocratic view. Allowance is made for interaction, and thus communication, on the strength of the argument that two or more people know more than one. However, interaction focuses on the adoption of the plan. Once adopted, the plan forms an unambiguous guide to action, and no more discussion is envisaged. Adoption also implies a definite image of the future. Consideration of time is restricted to the phasing of works.

In the 'context of accommodation' there is awareness of divergent perceptions of various parties. As we know, the codes used by the plan-maker and those responsible for subsequent decisions can also differ. Information can change perceptions, but is unlikely to bring the parties to full agreement. Communication in the context of accommodation may be envisaged precisely as the dual construction of texts by authors and readers that, following Eco (1979, 5-11), we have portrayed above.

Strategic plans (Figure 1.4) deal with the co-ordination of actors in a context of accommodation. Actors take decisions all the time, so co-ordination is continuous. Since all actors want to keep options open, timing is crucial. Rather than a finished product, a strategic plan is a momentary record of fleeting agreements, forms a framework for negotiations, and is indicative. The future remains open. Action never flows automatically from the plan. Each decision needs justification in its own right. The plan needs to be interpreted, much as judges interpret (and thereby change!) the law, where strict adherence would create anomalies.

In all this we must bear in mind that, in a context of accommodation, planning is more than the making of plans. Next to visible ones, like plans, planning also generates invisible products (Friend, Hickling 1988, 101-105). The shared views amongst participants are examples of invisible products of communicative planning. Invisible products may perform, even before the plan is finished. 'Pre-performance' of a plan (Wallagh 1992, 101-102) in fact refers to invisible products playing a useful role, even before the visible ones become available.

Figure 1.4

	Project plans	Strategic plans
object	material	decisions
interaction	until adoption	continuous
future	closed	open
time element	limited to phasing	central to problem
form	blueprint	minutes of last meeting
effect	determinate	frames of reference

The general preference will be for a 'context of accommodation' and appropriate forms of strategic, and hence communicative planning. So, before we go on to explore the implications for evaluation, let us hasten to add that in some situations, it is appropriate, nevertheless, to think of planning as the making of schemes that are subsequently adhered to. In other words, some planning may, indeed, be what the classic view shows; some communication is more usefully seen as a flow that wants to be as undistorted as possible; in some instances, we had better talk in terms of a 'context of command'; in some instances, 'technocracy', with its characteristic patterns of decision-making and action, is more appropriate than 'sociocracy'. It is just that such project planning seems less interesting, not least from the point of view of evaluation. On the other hand, the form of strategic planning that we have been discussing is extremely challenging.

1.5 Conformance versus performance

A project plan within the context of command, taking a technocratic approach and invoking the flow model of communication, is expected to lead to specified results. Evaluation can follow means-ends logic, measuring conformity of outcomes with the plan. Technically, this can be complex, but the logic is simple. Not so with evaluating strategic plans, where departures do not necessarily indicate failure. All that we can insist on is that planning contributes to the goal of soundly-justified decisions. In this section we describe how the evaluation of such strategic plans must proceed.

We start with planning as a communication process according to the second model. Strategic plans are evidently communicative. By the very nature of communication as the dual construction of texts, 'success' of communication cannot, however, be measured invoking conformance as a criterion. The evaluation criterion of planning according to the second model is performance (Williams 1976, 278). It does not presuppose conformance of actions with the original plan (Fudge, Barrett 1981). After all, knowing that a decision conforms to a plan says nothing about how soundly justified the decision was. What decision takers and their critics will rather want to know is whether the decision is adequate to the situation.

The prototype performance study concerns the 1934 General Expansion Plan of Amsterdam, the first Dutch plan based on extensive surveys (Postuma 1987), that boldly reached forward to the year 2000. The surveys analysed housing, land acquisition and compulsory purchase, in each instance establishing the role of the plan. Before the war housing schemes simply followed the plan. However, the Port Authority took no notice of it, so on this count it failed. After the war, sand was difficult to obtain, and so it was decided to raise the land by less than the plan had intended. Since Amsterdam is of course located below sea level in the Western Netherlands, its rainwater storage area needed to be enlarged. Now although the pre-war plan was not followed, it did inform decision-making. In using it, the Burgomaster and Aldermen (the mayor and executive council) pointed out where and why they had not followed it. The council, too, referred to it. So the idea of a plan 'working' by assisting decision takers without necessarily being followed makes sense.

Various publications have elaborated on this. Research usually starts with an analysis of the plan, identifying its intended performance. The set of decisions addressed in the plan forms the research population. Next, these decisions are analysed so as to assess the performance of the plan. Although not a criterion of evaluation, conformance has a practical role in identifying the cases needing further attention (Mastop, Faludi 1993). If only they can be attributed to the plan, decisions conforming to it are deemed to be instances of performance, but other decisions get more attention. If the plan did play a role in the consideration of alternatives, it may be said to have performed usefully. This last difference in the analysis of decisions that conform with the plan or not is more a question of efficiency -- it saves time -- than of methodological rigour, and doing away with it cannot be seen as a contribution by the present paper to the methodology of performance evaluation.

Here we focus on new aspects. In fact, we have already indicated what we mean by plan performance: the role plans play in the justification of decisions. We have also referred to the idea that justification involves actors --- critics -- besides the decision takers. Decision makers and critics together form the arena of justification. Another point already mentioned is that the plan cannot be the sole means of identifying the research population. After all, no plan maker can identify in advance all the decisions in which the plan will play a part.

Conceiving the performance of plans as entailing communication between their authors and those concerned with subsequent decisions (the decision takers and their critics) has another important implication (Figure 1.5). It is that a plan as such has no direct impact on the justification of decisions. Rather, what has an impact (what 'performs') is the reconstruction of the plan available to the decision taker and the critics; this being our third addition to current thinking about performance research. It is this reconstruction that may form part of their frame of reference, depending of course, on the codes which they use.

Figure 1.5

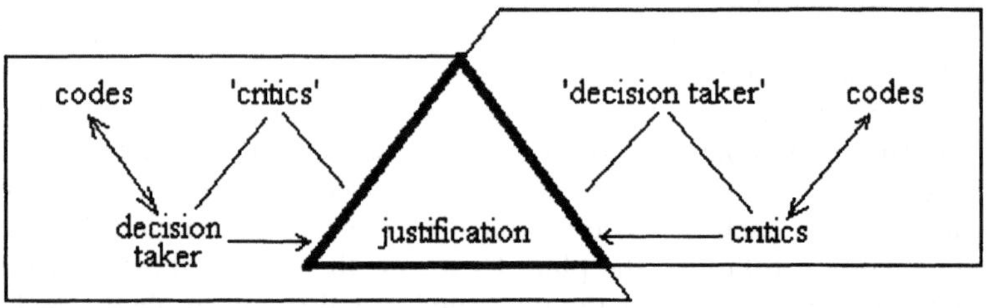

So planning is a systematic attempt to construct frames for the justification of decisions, the specific contribution being to put them in their wider field of choice. The point we have made is that the plan maker offers this frame to others, and they actually invoke the plan. To paraphrase Argyris and Schön (1981), it is not the 'espoused plan', but the 'plan in use' that performs (Faludi, Van der Valk, 1994). This is not the written or printed plan, but what is reconstructed every time decision takers and their critics refer to it. After all, virtually no one reads the whole plan before taking a particular decision. The most anyone should expect is that bits and pieces are read. It is equally obvious that plans can be read in different ways (Mandelbaum 1990). This 'plan in use' is thus the plan present in the codes used in justifying decisions. An espoused plan must become a 'plan-in-use' (or at least a part of it) presented in the codes with which the decision takers and their critics shape the justification of decisions. Here a problem arises that adds to the uncertainty of planning. A plan never automatically becomes part of the frame of reference, or the codes used, in justifying decisions. First the plan must succeed in penetrating this frame. Here, competition with other plans or the secret agendas of other actors can be stiff.

The evaluation of planning in the sense of the activity evolving around plans is often neglected by authors whose primary interest is the day-to-day work of planners. Planners communicate, use arguments and strategies. In everyday practice there is a plan in use, but no espoused plan (Figure 1.5). The espoused plan plays no direct role in the justification of decisions.

A fine example of the distinction between the plan in use and the espoused plan can be found again in the 1934 General Extension Plan of Amsterdam that is still being used to justify decisions about the development of Amsterdam (De Groot 1993, 15; Lambert 1990). Thus, arguments are being advanced as to whether certain developments do, or do not, complement the characteristic 'lobe' model imputed to that plan, which for long considered paradigmatic for Dutch planning. In fact though, the lobe model was articulated in the fifties, and it deliberately contravened the General Extension Plan's earlier ideas on concentrated development (Wallagh 1990, 96-97, 116-117). So the General Extension Plan (being the plan in use) is still a part of the codes that justify present-day decisions on a subject that was never in the espoused version of the General Extension Plan.

To formulate a plan, a planner needs to have a notion of its users. "The author has thus to foresee a model of the possible reader (..) supposedly able to deal interpretively with expressions in the same way as the author deals generatively with them." (Eco 1979, 7) At the same time a plan (just like every other text) aims to develop the competence of others in using the codes it presents, for example, by introducing them to new planning concepts that should cause them to perceive problems and their solutions in a particular way. Thus a plan impinges upon the frame of reference of its model users ('decision taker' and 'critics' in Figure 1.6). However, this model user is nothing but a construct of the planning subject. The actual user may have a frame of reference other than that imputed to the 'model user'.

Figure 1.6

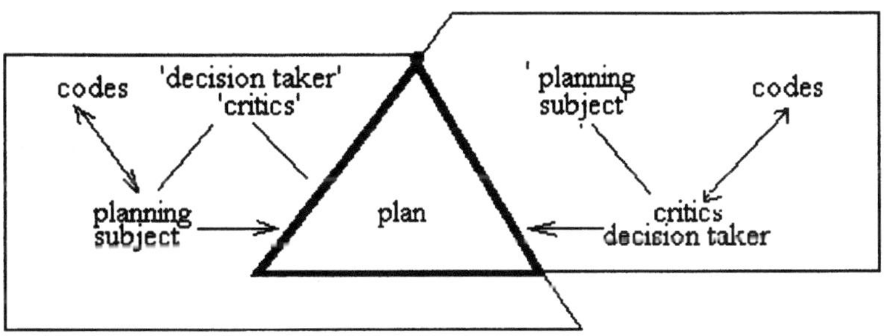

Thus, in the context of accommodation, a plan and its model user are both 'open'. Unlike in the context of command in which the actual users of a plan must conform to the 'model user', a plan can have meaning, even if the actual users do not conform to the 'model user'. To repeat, planning in a context of accommodation is planning under uncertainty. A plan maker can, indeed, accommodate to the frame of reference (the codes) of the target group of the plan. Such a strategy requires identifying the addressees of the plan, and ideas as to how to reach them by anticipating the contexts of the justifications of the decisions involved. Note, however, that during plan preparation only the potential users and not the actual users are known.

Since planning is something done under uncertain conditions, the precise circumstances under which follow-on decisions will be taken are unknown to the plan-maker, who may not know the identities of the relevant decision takers. This applies even more so to the 'critics' inhabiting the arenas in which decisions are justified. Clearly, the world is in flux.

Evaluation of the effects of a plan on the quality of the argument adduced in favour of a decision (its justification) -- what we call plan performance -- demands a research design of its own.

Research into performance starts by defining the research population, for it cannot be identified by simply looking at the plan. This fourth addition to current thinking about performance research is a consequence of the fact that a plan sometimes influences decisions which the plan maker did not, indeed could not, anticipate. Decisions relevant to the plan and its purpose may be taken by actors far different from the model users whom the plan maker had in mind. An example is the Dutch Fourth National Physical

Planning Report. After its publication, the Minister of Finance created a fund for infrastructure and, in selecting projects for funding, took account of the Report, which thus performed a useful role, without actually having addressed the issue of the distribution of these moneys. After all, at the time when the Fourth Report was made, no such fund existed. The fact that the Fourth Report did provide arguments for its use is to its credit even so.

In this, plans are analogous to scientific theories. In both cases, model uses are imagined. "But while theories fix their own intended interpretations, they do not fix their own domains of application, nor the resources for detection of entities they posit. ... New applications may arise with changes in collateral knowledge. Indeed, it is a measure of a theory's success when posited entities acquire a technological role, and applications for which the theory was not designed become possible." (Laudan, Leplin 1991, 456)

So it is obvious that, if we were to restrict the research population to decisions anticipated in the plan, such successes would go unnoticed.

If we cannot glean the research population (the decisions for which the plan provides a framework) from the plan itself, what can we do? With the advantage of hindsight, we can only construct a research population by interpreting the putative meaning of the plan to those concerned with subsequent decisions. All those decisions that have a bearing on it should be investigated, irrespective of whether the maker of the plan has actually thought about them. (In terms of communication theory: irrespective of whether the plan maker has included those responsible in his conception of the model users of the plan.) In fact, the evaluator has to judge the plan as its maker(s) might do in the light of information that has become available since it was adopted.

So the research design proposes three steps. The first is to:

1. Identify the decisions that the plan should influence.

Of course this requires detailed analyses of what is going on after the plan has been adopted. The demands on the researchers increase still further because they then need to establish the arenas within which these decisions are, or must be justified. Decisions are of the nature of commitments that mean that revoking decisions entails costs. Because such costs are large, the justification of decisions implying the strongest commitments should get priority. So we must ask what it costs to go back on decisions and whether the decisions are legally binding. May other actors call the decision takers to account if they rescind their decisions? And if so, who are these actors? Can they go to court, that is "mobilize the state" (Moore 1973)? What other sanctions do these 'critics' have? Then the second step is to:

2. Decisions imply commitments: identify them and the arenas where they are justified, and also identify the critics who operate in these arenas.

We turn to the actual justification of decisions and the role of the espoused plan in shaping the 'codes' used. As we have seen above, an espoused plan can only perform through the codes that shape the justification of decisions. It is the interpretation of the plan -- the 'plan-in-use' available in the codes of decision taker and critics -- that plays a role in the justification of decisions. The last question of the research design, and at the same time the answer to whether a plan has performed, is therefore:

3. Has the whole plan, or a part of it, helped in shaping the codes used in justifying subsequent decisions, and has this improved the quality of the justification of decisions in terms of taking account of the wider field of choice?

We deliberately speak of improving justifications. Plans can be more of a hinder than a help in justifying decisions. Thus, a plan based on outdated assumptions can be worse than no plan at all. Critics may also pin decision takers down to certain lines of action, making use of a plan which is no longer appropriate. For example, towards the end of the 1980s, the Dutch National Physical Planning Agency wanted to play down the concept of 'Randstad', but largely because planners had used this concept in the past, it had taken root. Irrespective of whether the agency was right in trying to do away with it, this shows that plans do influence decision situations in ways which the plan makers themselves may dislike.

Answering the last question it is important to realize that codes evolve. They can therefore highlight differences between various critics of the decision (Peters, Marshall 1993). The present research design is no broom for sweeping such differences under anyone's carpet.

1.6 Conclusions

The research design provides reasoned arguments about the way planning improves the justification of decisions. However, under sociocratic planning, evaluators (like planners) are not objective, technical experts. At the end of the day, there is no irreversible 'thumbs up' or 'thumbs down'. However, evaluation allows us to take a stand as regards the performance of a plan that is supported by (necessarily fallible) arguments. Its results may not be unambiguous. Some aspects of planning may have worked, and some not. Anyhow, such findings allow us to have rational discussions about the success of planning and how to improve it, which is perhaps all we can ask.

This performance evaluation is not 'theory driven' (Chen 1990): its purpose is not to test hypotheses behind the plan. Rather, the relation between evaluation research and improvements in conducting planning work is described as 'methodological reflection'. Thus empirical research ".. does not so much provide the *basis* as the *occasion* for methodological argument." (Faludi, Mastop 1982, 254)

Methodological reflection proceeds by way of logical analysis. Empirical research gives signals as to what to focus on.

Such methodological reflection can lead to strategies for improving performance. The ideas on 'planning doctrine', for example, originated from reflection on a case (described in Faludi, 1987, 128-132) in which, despite poor planning methodology, planning seemed to perform alright. In this way planning methodology has an empirical component, too.

Anyhow, what we have demonstrated is the fertility of planning as a systematic attempt to construct frames for the justification of decisions. The aim of planning in a context of accommodation is not conformance between the plan and the action that follows (let alone its material effects). This aim is rather to improve the justification of decisions. The evaluation of performance needs a research design that focuses on the arguments advanced during this justification.

This research design needs to pay regard to the idea that justification takes place in an arena, with critics discussing the merits of decisions. These critics help to shape the definition of the decision situation. The researcher must be aware that it is not the plan, but its reconstruction by the decision taker and its critics, that performs during the justification of decisions. Finally, the research population must not be restricted to decisions stated in the plan. Rather, it must be established independently.

Maybe our proposal of a different perspective on the role of evaluation in improving planning practice is even more important. Evaluation is not employed to falsify poor policy theories but to provide occasions for methodological reflection on the function of communicative planning as a means of improving the justification of decisions.

References

Alexander, E.R. (1993) "The concept of planning doctrine", in A. Faludi (ed.) Dutch Strategic Planning in International Perspective, SISWO, Amsterdam, 23-30.

Alexander, E.R. and A. Faludi (1989) "Planning and plan implementation: Notes on evaluation criteria", Environment and Planning B: Planning and Design 16, 127-140.

Alexander, E.R. and A. Faludi (1990) "Planning doctrine: Its uses and implications", Werkstukken 120, Planologisch en Demografisch Instituut, Universiteit van Amsterdam

Argyris, C. and D.A. Schön (1981) Theory in Practice: Increasing Professional Effectiveness, Jossey-Bass, San Francisco.

Bardach, E. (1980) "On designing implementable programs", in G. Majone and E. S. Quade (eds.) Pitfalls of Analysis, John Wiley, Chichester, 138-158.

Centre for Environmental Studies (1970) The LOGIMP experiment, Information Paper 25, London.

Chen, Huey-Tsyh (1990) Theory-Driven Evaluations, SAGE, Newbury Park.

Cronbach, L.J., S. Robinson Ambron, S.M. Dornbusch, R.D. Hess, R.C. Hornik, D.C. Phillips, D.F. Walker and S.S. Weiner (1980) Toward Reform of Program Evaluation: Aims, Methods, and Institutional Arrangements, Jossey-Bass, San Francisco/London.

Eco, U. (1979) The Role of the Reader: Explorations in the Semiotics of Texts, Indiana University Press, Bloomington/London.

Edwards, A.R. (1991) "Planning betwist: Naar een rehabilitatie van de communicatieve planningconceptie", in: P. de Jong and A.F.A. Korsten (eds.) Planning, Praktijk en Onderzoek, SISWO, Amsterdam, 59-63.

Faludi, A. (1984, first published 1973) Planning Theory, Pergamon, Oxford.

Faludi, A. (1987) A Decision-centred View of Environmental Planning, Pergamon, Oxford.

Faludi, A. (1989a) "Conformance vs. performance: Implications for evaluations", Impact Assessment Buletin 7, 135-151.

Faludi, A. (1989b) "Perspectives on planning doctrine", Built Environment 15/1, 57-64.

Faludi, A. and J.M. Mastop (1982) "The 'IOR school': The development of a planning methodology", Environment and Planning B: Planning and Design 9, 241-256.

Faludi, A. and A. van der Valk (1994) Rule and Order: Dutch Planning Doctrine in the Twentieth Century, Kluwer Academic Publishers, Dordrecht.

Foley, D.L. (1963) Controlling London's Growth: Planning the Great Wen 1940-1960, University of California Press, Berkeley.

Foley, D.L. (1964) "An approach to metropolitan spatial structure", in M.M. Webber et al. Explorations into Urban Structure, University of Pennsylvania Press, Philadelphia, 21-78.

Forester, J. (1980) "Critical theory and planning practice", Journal of the American Planning Association 46, 275-285.

Forester, J. (1987) "Anticipating implementation: Normative practices in planning and policy analysis", in Frank Fischer and John Forester (eds.) Confronting Values in Policy Analysis, SAGE, Newbury Park, 153-173.

Forester, J. (1989) Planning in the Face of Power, University of California Press, Berkeley.

Friedmann, J. (1965) "Venezuela: From doctrine to dialogue", National Planning Series 1, Syracuse University Press, Syracuse.

Friedmann, J. (1969) "Notes on societal action", Journal of the American Institute of Planners 35, 311-318.

Friedmann, J. (1973) Retracking America: A Theory of Transactive Planning, Doubleday Anchor, Garden City NY.

Friend, J. and A. Hickling (1988) Planning Under Pressure: The Strategic Choice Approach, Pergamon, Oxford.

Friend, J.K. and W.N. Jessop (1977) Local Government and Strategic Choice, Pergamon, Oxford.

Fudge, C. and S. Barrett (1981) "Reconstructing the field of analysis", in S. Barrett and C. Fudge (eds.) Policy & Action: Essays on the Implementation of Public Policy, Methuen, London/New York, 249-278.

Groot, A. de (1993) "Een probleemverkenning aan de Westrand van Amsterdam", in Verslag van het Symposium dd. 18-6-1993 over het AUP en de Actualiteit van het Regionale Ontwerp, Amsterdamse Raad voor de Stadsontwikkeling, Amsterdam, 15-18.

Gunsteren, H.R. van (1976) The Quest for Control: A Critique of the Rational-central-rule Approach in Public Affairs, John Wiley, London

Habermas, J. (1984) The Theory of Communicative Action, Polity Press, London

Healey, P. (1992) "A planner's day", Journal of the American Planning Association 58, 9-20.

Healey, P. (1993) "The communicative work of development plans", Environment and Planning B: Planning and Design 20, 83-104.

Korthals Altes, W. K. (1992) "How do planning doctrines function in a changing environment?", Planning Theory Newsletter 7-8, 100-115.

Korthals Altes, W.K. (1993) "Doorwerking als communicatie", Bestuurswetenschappen 47(5), 319-337.

Kreuz, R.J. and R.M. Roberts (1993) "The empirical study of figurative language in literature", Poetics 22, 151-169.

Krumholz, N. and J. Forester (1990) Making Equity Planning Work, Temple University Press, Philadelphia.

Lambert, D. (1990) "Speculaties over de kwaliteit van de toekomstige stad", in Donald Lambert, Arno de Vries (eds.) Stadsontwerp Amsterdam: Stedebouwkundig Ontwerpen, Publikatieburo Faculteit der Bouwkunde, Delft, 9-21.

Laudan, L. and J. Leplin (1991) "Empirical equivalance and underdetermination", The Journal of Philosophy 88 (9), 449-472.

Loon, J.F.G van (1958) Rules and commands, Mind 67, 514-521.

Mandelbaum, S.J. (1990) "Reading plans", Journal of the American Planning Association 56, 350-356.

Mastop, J.M. (1987, first published 1984) "Besluitvorming, handelen en normeren", Planologische Studies 4, Planologisch en Demografisch Instituut, Universiteit van Amsterdam.

Mastop, J.M. and A. Faludi (1993) "Doorwerking van strategisch beleid in dagelijkse beleidsvoering", Beleidswetenschap 7(1), 71-90.

Moore, S.F. (1973) "Law and social change", Law & Society Review 7, 719-746.

Peters, M. and J. Marshall (1993) “Educational Policy Analysis and the Politics of Interpretation”, Evaluation Review 17 (3), 310-330.

Pitkin, H.F. (1972) Wittgenstein and Justice: On the significance of Ludwig Wittgenstein for Social and Political Thought, University of California Press, Berkeley

Postuma, R. (1987) “Werken met het AUP: Stadsuitbreiding van Amsterdam 1939-1955”, Working Papers 97, Institute of Planning and Demography, University of Amsterdam.

Throgmorton, J.A. (1993) “Planning as a rhetorical activity: Survey research as a trope in arguments about electric power planning in Chicago”, Journal of the American Planning Association 59, 334-346.

Valk, A. van der (1989) “Amsterdam in aanleg: planvorming en dagelijks handelen 1850-1900”, Planologische Studies 8, Planologische en Demografisch Instituut, Amsterdam.

Wallagh, G. (1990) “Naar een nieuwe tijd - Strategische ruimtelijke planning in Amsterdam en omgeving 1955-1969”, Working Papers 126, Institute of Planning and Demography, University of Amsterdam.

Wallagh, G. (1992) “Het structuurplan, toepassing en mogelijkheden verkend”, Working Papers 149, Institute of Planning and Demography, University of Amsterdam.

Williams, W. (1976) “Implementation analysis and assessment”, in Walter Williams and Richard F. Elmore (eds.) Social Program Implementation, Academic Press, New York, 276-292.

Winch, P. (1973 first published 1958) The Idea of a Social Science and its Relation to Philosophy, Routledge & Kegan Paul, London

Witteveen, W.J. (1988) “The rhetorical labors of Hercules”, Working Paper 15, Leyden Institute for Law and Public Policy, Leiden

Witteveen, W.J. (1992) “De wet als boodschap, de wet als tekst”, in W.J. Witteveen, H.D. Stout and M.J. Trappenburg (eds.) Het Bereik van de Wet, Tjeenk Willink, Zwolle, 13-42.

Zonneveld, W. (1991) “Conceptvorming in de ruimtelijke planning: patronen en processen”, Planologische Studies 9A, Planologisch en Demografisch Instituut Universiteit van Amsterdam

Notes

[1] This paper has been published previously, in a modified version, under the title of "Evaluating communicative planning: a revised design for performance research", in European Planning Studies, Vol. 4, pp. 403-418.

[2]In fact, the case of arguing about a decision with ourselves can also be seen as a situation in which we meet an imaginary other.
[3]Hopefully, but not necessarily, decisions are justified in the light of that plan. We say hopefully because, in planning as elsewhere, prior investments do not always bear fruit.
[4]The decision-centred view argues that, in the first instance, planning is about arranging the decisions which the planning subject has the competence to take. Out of these decisions, meaningful packages must be formed. However, for the present purpose this is neither here nor there.
[5]Being a planning input, it will of course concern what has been called above the wider field of choice.
[6]We do not want to go further into this, but analysing the role of planning doctrine -- a concept developed to analyse the influence of a durable coherent body of thought on planning (Faludi, 1989b, 60-61; Alexander, 1993; Alexander, Faludi, 1990; Korthals Altes, 1992) -- as a long-term structure underlying communication codes in planning performance may be a help in analysing this learning capacity. This analysis may be a part of the research design below.

Andreas Faludi
Willem Korthals Altes
Planologisch en demografisch Instituut
Faculteit der Ruimtelijke Wetenschappen
Universiteit van Amsterdam
Nieuwe Prinsengracht 130
1018 VZ Amsterdam
The Netherlands

2 ON THE ROLE OF WILL-SHAPING IN PLANNING EVALUATION

Henk Voogd

2.1 Introduction

Plan and project evaluation have become important activities in modern public planning and administration. Especially in the field of environmental and infrastructure planning, many examples can be found of a systematic assessment and appraisal of alternative policy proposals, also called ex ante evaluation (e.g. see Voogd, 1994). Selective overviews of evaluation approaches can be found in the following books: Cochrane and Zeleny (1973), Lichfield, Kettle and Whitbread (1975), Nijkamp (1980), Kmietowicz and Pearman (1981), Voogd (1983), Shofield (1987), Nijkamp and Voogd (1989), Shefer and Voogd (1990), Nijkamp, Rietveld and Voogd (1990).

Of course, the ultimate choice of the preferred evaluation approach is always a function of the nature of the problem, the interested parties and the planning context. A recent analysis of recent infrastructure planning processes in the Netherlands has found that many problems during the evaluation phases of a planning process relate to the *interaction* between the planners and the planning environment (see Alteren et al., 1990). In particular the inadequate integration of evaluation and planning processes is seen as a major reason for discontent. The application of methods often neglected the need for evaluation to accompanied with a 'will-shaping process' to familiarize people with the alternatives and the evaluation outcomes.

In this paper, we focus our attention on the relationship between evaluation and planning processes. In the next section we show that evaluation may not be simplified to the mere application of one or two methods, however useful these methods may be. Evaluation in urban and regional planning can be seen as a complex set of social activities directed towards the specification and choice of a set of goals and related plans and/or projects (e.g. see Archibugi, 1994). Will-shaping is an important part of it, and this can be realized in several ways. Some theoretical foundations of the 'will-shaping' concept are discussed in more detail in section three. Section four is devoted to some consequences

D. Borri et al. (eds.), Evaluating Theory-Practice and Urban-Rural Interplay in Planning, 23–34.

of this concept for the formal structuring of the evaluation and planning process. We conclude with some summaries.

2.2 Characteristics of planning evaluation

An analysis of infrastructure planning processes in the Netherlands has revealed that a major fundamental reason for problems is that many people, including many 'technical' planners, have a limited notion of what planning evaluation is (e.g. see Alteren et al., 1990). They consider an evaluation process to be mainly an 'intellectual process'. In this view, methods and techniques play a role that is next in importance to disciplinary-based empirical knowledge. However, there are at least three more dimensions of an evaluation process that are all extremely relevant to understanding the practical problems that arise in applying evaluation methods. These are summarized in Figure 2.1.

Figure 2.1: Four dimensions of an evaluation process

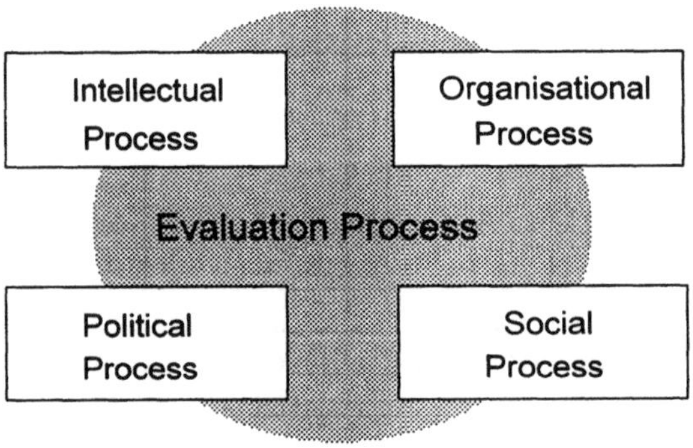

Planning evaluation is not only an intellectual process, in which the proper alternative options are specified and their various impacts are compared. It is also a 'political process', in which power structures, consultation, mediation and negotiation play a prominent role (see Forester, 1989). An important consideration in this process concerns the question, 'Who gains and who loses?'. The answer to this question is strongly related to the problem of 'equity' versus 'efficiency' (e.g. see Miller, 1985). In practice, defining what equity and efficiency involve is very problematic. The reason is that such definitions may have many political implications. An assessment of the distributional effects of planning proposals implies that explicit consideration must be given to social categories (for instance, the definition of groups of actors involved in, or affected by, policy). The PBS framework of Lichfield that distinguishes between producers and operators on the one hand, and consumers on the other, is especially helpful here, for it explicitly considers the 'intellectual process' (e.g. see Lichfield, 1985, 1990). Practice, however, teaches that explicitly to classify groups may also be very difficult, especially when the evaluation is done in relative isolation and, therefore, without close consultation with the groups concerned. When outcomes do not reflect the preferences of a group, it is certain to voice criticisms. Consequently, close cooperation with the interests concerned is strongly recommended, which implies explicit attention to evaluation as a 'political process'.

However, planning evaluation must also be considered as a 'social process'. It is particularly important for planners concerned with the development and evaluation of planning proposals to understand the social context of public decision making. There is a vast literature on individual decision making, much of which is directly about evaluation as an intellectual process. In the use of planning evaluation methods an important point is that politicians and interest groups make choices based on their *perceptions* of reality that often differ from reality as defined or seen by planners. To effect the desired changes, those engaged in public planning need to understand the circumstances and criteria that people employ in arriving at a judgment relating to a proposed plan or project. It is also necessary to gain knowledge on how people evaluate information in their opinion-making process. Evidently, many outside-participants in public planning processes do not successfully obtain relevant information, because much that is important never reaches the pages of a project plan. Therefore, in an early stage of the evaluation process selecting the most effective means of communication with various social and political groups will always remain important in every public evaluation process with conflicting issues (this corresponds to concepts in planning theory which focus on planning as a 'communication structure', e.g. see Habermas, 1973; Van Gunsteren, 1976; Teisman, 1992).

The four dimensions of an evaluation process, as we have mentioned, to some extent explain the difficulties encountered when evaluation methods are applied in planning practice. Many decision-making processes need *time*. Time that can be used for a proper analysis of the consequences, which will also always increase uncertainties due to the dynamics of political viewpoints, of value systems, of society, to decrease uncertainties. In other words, time needed for *will-shaping*, to familiarize with problems and with solutions. By will-shaping we mean forming a mental faculty common to many people through which they deliberately choose or decide upon a course of action.

2.3 Some theoretical observations

Resembling a negotiation process, will-shaping can be seen as a process that aims to synchronize attitudes towards and preferences for certain goals. Negotiation techniques, as much about the process of decision-making as the decision itself, include procedures and ground rules for negotiations between different parties that aim at solving problems by fostering consensus between the various interest groups - reconciling different calculations of costs and benefits (Susskind and Cruikshank, 1987; Susskind, 1994). Will-shaping is essential in arriving at a consensus. It will be part of a negotiation or mediation process, but it can also be witnessed in any - less formally structured - planning or decision-making process. For example, promotional activities around urban revitalisation projects are often intended to create a permissive social attitude towards them (Ashworth and Voogd, 1990).

The relationships between attitudes, preferences and human behaviour have been widely studied, in particular in cognitive psychology (Fishbein and Ajzen, 1975; Dawes and Smith, 1985, Greenwald, 1989). The 'theory of planned behaviour' (Ajzen, 1988, is very interesting in stating that a choice is made between alternative behavioural patterns on the basis of attitudes towards these patterns. Bentler and Speckart (1981) have shown that habit may be as important as intentions in determining behaviour. If behaviour becomes habitual, attitudes may even change without accompanying changes in behavioural patterns. However, in general attitudes and preferences are seen as determinants of behaviour and hence of decision-making (see also Dwyer et al, 1993). Obviously, attitudes are more enduring dispositions than preferences (Ajzen, 1987). Changes in attitudes are likely to underlie more permanent behavioural changes. Such changes can probably be accomplished by directly changing attitudes, or, at least in certain circumstances, by indirectly changing behaviour.

The process of will-shaping has not been much well studied in public decision-making processes, although it is a crucial cognitive characteristic of human beings (e.g. see Grossberg, 1982; Kolb and Wishaw, 1990). For a proper understanding of will-shaping it is essential to divide the process of the forming of a common mental faculty into three distinct mental processes concerning information: its projection, transmission and reception. *Its projection* concerns the mental representation of the problem and the suggested solutions (i.e. the alternatives) by the users of evaluation research results (viz. the public and politicians). Different media can be used for this *transmission*. For example, between planners and political executives there is usually direct communication via meetings and/or planning reports. Transmission may also occur through intermediaries like news media, and citizens' participation. However, projected information usually has to compete with message interference (e.g. unexpected events, past experiences) from other sources of information. Consequently, the image of the planning problem and planning alternatives received by the audience can be quite different from the one intended by the planners. In other words, it is essential to distinguish *information reception* as a specific component of the will-shaping process, because this component ultimately determines the success of the evaluation process in the sense that the receiver (i.e. the audience) returns a useful reaction to the sender (i.e. planning authorities). It is visualized in Figure 2.2.

Figure 2.2: Information processing tasks in a will-shaping process

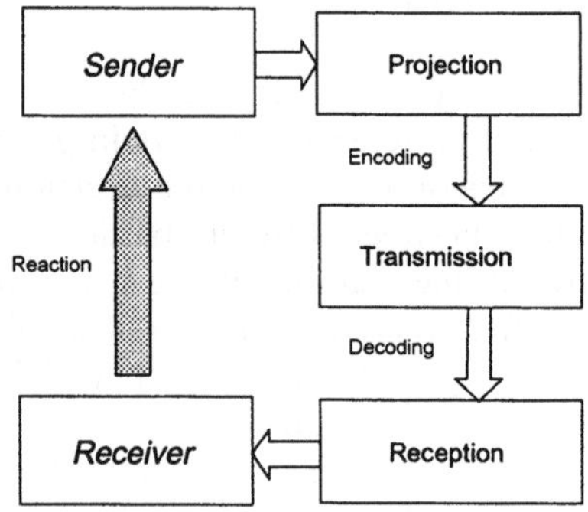

Figure 2.2 is just one way to represent human information processing. In cognitive psychology at least three different types of models of human information processing have been formulated (e.g. see Miller, 1988; Sanders, 1990). *Discrete-serial stage models*

assume the existence of serially organized stages of information processing, which each transmits output in one final step only after which a subsequent stage can become active. *Continuous models* assume that information processes, organized in parallel, continuously transmit preliminary results of their transformation to contingent processes. Finally, *hybrid models* are recognized, which form a mixture of both discrete and continuous models.

Public planning and promotion have a lot in common (e.g. see Ashworth and Voogd, 1990). Evidently, the will-shaping process may be structured by explicit promotional activities of evaluation results. Promotion tries to evoke a specific change in an audience's attitude or behaviour. The change sought is a specific response from the target group. From persuasion theory it is known three different forms of response are possible (see Roloff and Miller, 1980). Firstly, promotion may be *response shaping*. This is similar to learning: evaluation activities may attempt to shape the response of an audience by 'teaching' it how to behave and offer positive reinforcement for learning. If audience responses favourable to the planner's purpose are reinforced by rewards to the audience, positive attitudes are developed toward what is learned. The audience fulfils a need for positive reinforcement, and the planner fulfils a need for a desired response from the audience at hand. Secondly, there is *response reinforcing*. If target persons in the audience already have positive attitudes toward the proposed alternatives, whether specific or general, the planner reminds them about the positive aspects of the solutions and stimulates them to feel even more strongly by displaying their attitudes through specified forms of behaviour. Many public policy activities in today's society are response reinforcing (EC fund raising, seeking political support, investments, and so on), but the people from the target group (e.g. EC authorities, investors, social groups) have to be motivated to do these things. Thirdly, promotion may be *response changing*. This is the most difficult task because it involves asking people to switch from one perceived image of the planning problem and alternatives to another. People are reluctant to change; thus, to convince them to do so, the planner has to relate the change to something in which the target person already believes. In persuasion theory this is called an 'anchor' (Roloff and Miller, 1980), because it is already accepted by the target person and will be used to tie down new attitudes or opinions. An anchor is a starting point for change because it represents something that is already widely accepted. Anchors can be beliefs, values, group norms, etc..

2.4 Structuring the evaluation process

It is well-recognized that planning evaluation is a continuous activity, which takes place in many stages of a planning process (e.g. see Lichfield et al. 1975). However, in planning practice the role of explicit ex ante evaluation, and therefore the role of systematic evaluation methods, is usually restricted to a stage in which several alternative proposals are available and subject to internal (i.e. within governmental organizations) discussions. Such an evaluation process will generally have the following simple structure: see Figure 2.3.

It starts with an analysis of the planning situation, which involves an analysis of the problems involved, the goals and objectives and the various relevant interest and/or target groups. Often simultaneously, work is done to find and design alternative options and to assess their impacts and other relevant characteristics. In the evaluation phase, work usually aims at constructing one or more rankings of the alternatives: the higher an alternative is ranked, the more it is recommended (at least, from the point of view of the weight of the criterion in question if a multicriteria evaluation has been performed).

Figure 2.3: A simple structure of basic activities in an evaluation process.

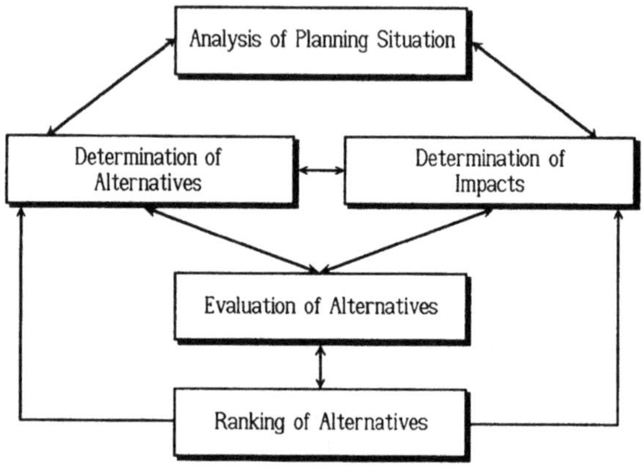

Figure 2.4: Example of a combination of an evaluation result (i.e. effectiveness score) to another criterion (i.e. urgency).

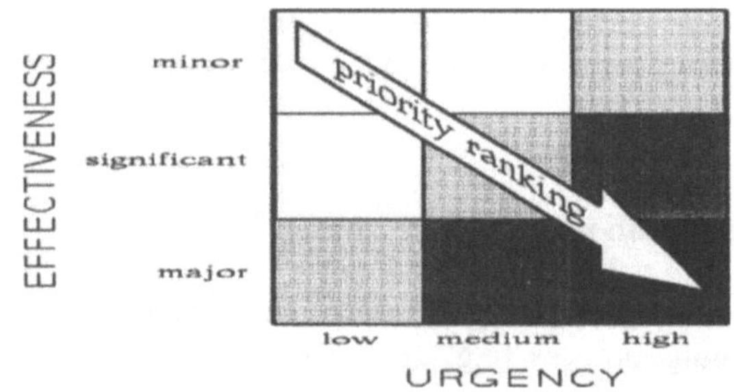

Practice teaches that evaluation results, derived according to the model outlined in Figure 2.3, may be very soon 'out of date'. This is especially the case if the external negotiation process and/or citizen participation rounds provide 'new' information and/or other political values may become apparent. If the evaluation results are already included in a formal plan or proposal, it is usually considered by political executives as 'undesirable' to modify the analysis. This may be interpreted by the political opposition as a token of a 'drifting policy'. Therefore, preference is usually given to adaptations like that of Figure 4, i.e. linking an 'old' evaluation outcome to a 'new' preference represented by some vague criterion such as 'urgency' or the like. Such adaptations may be considered as illustrations of an inadequate will-shaping process.

At least two fundamentally different avenues can be explored to improve the evaluation process in this respect, viz. the strategic model and what can be (cf. Voogd, 1993) called the elaboration model. These models are roughly outlined in Figures 2.5 and 2.6.

Figure 2.5: Strategic Model

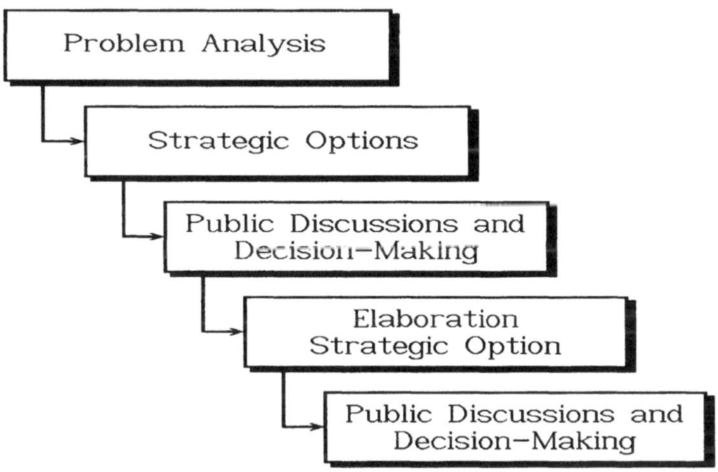

The *'strategic model'* is well-known in planning and management sciences (see, for instance, Capon et al., 1987; Sutherland, 1989). If applied in an evaluation context based on will-shaping, it starts with a general fundamental discussion on goals and objectives, resulting in 'strategic alternatives' included in a 'strategic plan'. These strategic alternatives are often built around metaphors like planning ideologies (Foley, 1960), planning concepts (Zonneveld, 1991) or planning doctrines (Alexander and Faludi, 1989).

The idea behind this strategic model is that the strategic plan should be the subject of public and political debate, so that the approved strategic plan can provide the constraints for a further elaboration of 'operational alternatives'. A weakness of the strategic model is that the strategic alternatives often have little or no appeal to the public. They are often too vague to be recognized as of vital importance to the future of their own living environment. The result is that the will-shaping process is hardly started in this phase. As such, many public objections may result in the next phase of operational planning because only then do the consequences of the strategic plan become evident to the public. It is usual then that in the phase of operational planning another fundamental discussion is started about the underlying principles and concepts of the strategic plan.

An alternative route, which takes into better account the will-shaping process, is the 'elaboration model'. This model starts immediately, of course, after a problem analysis, with the development of illustrative, but operational, alternatives. These alternatives have to cover a broad spectrum of principally different strategic directions, however, without being exhaustive within a strategic direction. These operational alternatives may be very useful in starting a public discussion and, consequently, a will-shaping process, in a very early phase of a planning process.

Figure 2.6: Elaboration Model

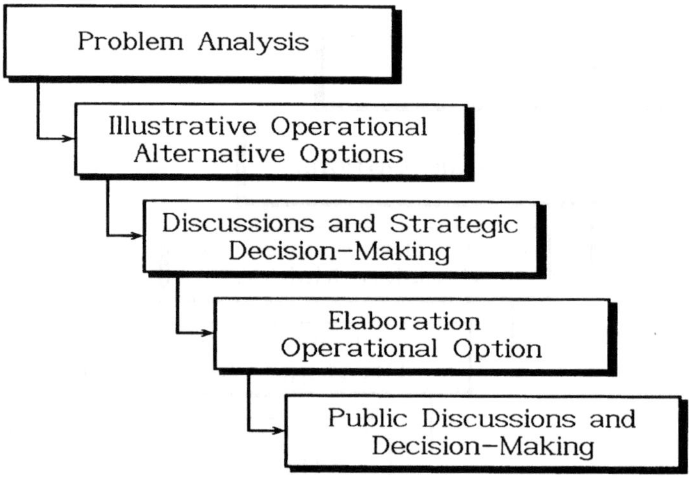

Both the strategic model and the elaboration model may have systematic evaluation activities. However, as clarified by Figure 1, evaluation is more than just a systematic comparison of alternatives to justify a preferred course of action. It also involves a broad orientation on the wishes and ideas of social groups, with the purpose for creating a 'power base', i.e. a broad consensus between many groups and participants in the planning and evaluation process. Only in this way can certain solutions not be overlooked and therefore not elaborated. A so-called 'open planning process', stressing communications with various interest and target groups, is much to be preferred in an early plan-making stage than an internal evaluation of the alternatives by means of a systematic evaluation method. Consequently, because the elaboration model stimulates the process of social 'will-shaping', it must be preferred to the strategic model.

2.5 Some concluding remarks

In this paper, we appeal for more attention to be given in evaluation to communication with social groups and their 'will-shaping'. In the light of earlier work by the author (e.g. Voogd, 1983, 1985) that aimed primarily at providing political executives with the best possible information and therefore improving public accountability, this may be seen as a change of mind. Indeed, in the present paper this target is to some extent weakened by stressing the necessity that an ex ante evaluation approach must also work towards a consensus. Obviously, the practical conduct of planners engaged in evaluation will always reflect both considerations. Ultimately, the balance of emphasis may depend in practice on the context and content of evaluation and on the phase of the decision-making process.

A key element with respect to the credibility and acceptability of ex ante evaluation is the way in which both methods and results are presented. The importance of good presentation not only holds for the accountability issue already mentioned but is also evident in an evaluation process that focuses on will-shaping. Thanks to modern technology e.g. computer graphics and so forth, many improvements are already on the horizon. However, a great deal of information resulting from evaluation cannot be represented graphically. More research is necessary with respect to possible 'interfaces' between evaluation and actual decision-making. Future research programmes should cover this important area, both empirical (case studies) and methodological.

It should be evident from this paper that there is no single recipe for evaluation. An important reason is the political and social dimensions of evaluation, but it would be unwise to conclude that therefore no rational evaluation approach needs to be pursued. On the contrary: the only way public authorities can legitimize their being towards citizens is by basing their work and working processes on scientifically rational points of departure! If, however, the factual behaviour of politicians and government is used as a behavioral norm and standard for evaluation, then this can only lead to a further deterioration of the situation. Evaluation will then be no more that a power play, where wishes are more important than impacts and political influence more important than arguments.

References

Ajzen, I. (1987) "Attitudes, traits, and actions: Dispositional prediction of behavior in personality and social psychology", in L. Berkowitz (ed.), Advances in Experimental Social Psychology, Academic, New York, 1-63

Ajzen, I. (1988) Attitudes, Personality, and Behavior, Open University Press, Milton Keynes

Alexander, E.R., A. Faludi (1990) "Planning doctrine: Its uses and applications, Planologisch en Demografisch Instituut, Universiteit van Amsterdam, Werkstukken PDI 120, Amsterdam

Alteren, S. van, B. van der Moolen, P. Ike, H. Voogd (1990) Integrale Projektstudies bij Infrastruktuurplanning (Integral Project Studies in Infrastructure Planning), Geo Pers, Groningen

Archibugi, F. (1994) "The disciplinary implications of environmental planning and evaluation", in H. Voogd (ed.) Issues in Environmental Planning, Pion, London, 173-186

Ashworth, G.J., H. Voogd (1990) Selling the City, Marketing Approaches in Public Sector Urban Planning, Belhaven, London

Bentler, P.M., G. Speckart (1981) "Attitudes cause behavior: A structural equation analysis", Journal of Personality and Social Psychology 40, 226-238

Capon, N., J.U. Farley, J.M. Hulbert (1987) Corporate Strategic Planning, Columbia University Press, New York

Cochrane, J.L., M. Zeleny (eds) (1973) Multiple Criteria Decision Making, University of South Carolina Press, Columbia

Dawes, R.M., T. Smith (1985) "Attitude and opinion measurement", in G. Lindzay, E. Aronson (eds), Handbook of Social Psychology, Random House, New York, 509-566

Dwyer, W.O., F.C. Leeming, M.K. Cobern, B.E. Porter, J.M. Jackson (1993) "Critical review of behavioral interventions to preserve the environment: Research since 1980", Environment and Behavior 25, 275-321

Faludi, A. (1987) A Decision-centred View of Environmental Planning, Pergamon, Oxford

Fishbein, M., I. Ajzen (1975) Belief, Attitude, Intention, and Behavior; An Introduction to Theory and Research, Addison-Wesley, Reading MA

Foley, D.L. (1960) "British town planning: One ideology or three?", British Journal of Sociology 11, 211-232.

Forester, J. (1989) Planning in the Face of Power, University of California Press, Berkeley.

Greenwald, A.G. (1989) "Why are attitudes important?" in A.R. Pratkanis, S.J. Breckler, A.J. Greenwald (eds), Attitude Structure and Function, Erlbaum, Hillsdale NJ, 1-10

Grossberg, S. (1982) Studies of Mind and Brain, Reidel Press, Boston

Gunsteren, H.R. van (1976) The Quest for Control, Wiley, New York

Habermas, J. (1973) Legitimationsprobleme im Spätkapitalismus (Problems of Legitimation in Late Capitalism), Suhrkamp, Frankfurt am Maim

Kmietowicz, Z.W., D.A. Pearman (1981) Decision Theory and Incomplete Knowledge, Gower, Aldershot

Kolb, B., Wishaw, I.Q. (1990), Fundamentals of Human Neuropsychology, Freeman, San Francisco

Lichfield, N., P. Kettle and M. Whitbread (1975), Evaluation in the Planning Process, Pergamon Press, Oxford.

Lichfield N. (1990) "Plan evaluation methodology: Comprehending the conclusions", in Shefer, D., H. Voogd (eds) Evaluation Methods for Urban and Regional Plans, Pion, London, 79-98

Miller, D.H. (1985), "Equity and efficiency effects of investment decisions: Multi-criteria methods for assessing distributional implications", in A. Faludi, H. Voogd (eds), Evaluation of Complex Policy Problems, Delftsche Uitgeversmaatschappij, Delft, 35-50

Miller, J.O. (1988), "Discrete and continuous models of human information processing: Theoretical distinctions and empirical results", Acta Psychologica 67, 191-257

Nijkamp, P. (1980) Environmental Policy Analysis, Wiley, Chichester.

Nijkamp, P., H. Voogd, L. Fusco Girard (1989) Conservazione e Sviluppo: La Valutazione nella Planificazione Fisica (Conservation and Evaluation Methods in Physical Planning), Franco Angeli, Milano

Nijkamp, P., P. Rietveld, H. Voogd (1990), Multicriteria Evaluation in Physical Planning, North Holland, Amsterdam/New York

Roloff, M.E. and G.R. Miller (eds) (1980), Persuasion: New Directions in Theory and Research, Sage, Beverly Hills

Sanders, A.F. (1990), "Issues and trends in the debate on discrete versus continuous processing of information", Acta Psychologica 74, 123-167

Shefer, D., H. Voogd (eds) (1989), Multicriteria Methods for Urban and Regional Plans, Pion Ltd, London

Shofield, J.A. (1987) Cost-Benefit Analysis in Urban & Regional Planning, Allen & Unwin, London

Susskind, L., J. Cruikshank (1987) Breaking the Impasse: Consensual Approaches to Resolving Public Disputes, Basic Books, New York

Susskind, L. (1994) Environmental Diplomacy: Negotiating more Effective Global Agreements, Oxford University Press, New York

Sutherland, J.W. (1989) Towards a Strategic Management and Decision Technology: Modern Approaches to Organizational Planning and Positioning, Kluwer, Dordrecht

Voogd, H. (1985), "Transportation policy analysis", Sistemi Urbani 3, 355-398

Voogd, H. (1993) Integrale Prioriteitstelling en Beleidsvorming Ministerie van Verkeer en Waterstaat, Project Integrale Prioriteitstelling Hoofd Infrastructuur, Den Haag

Voogd, H. (ed.) (1994) Issues in Environmental Planning, Pion, London

Zonneveld, W. (1991) "Conceptvorming in de ruimtelijke planning", Planologische Studies 9AB, Planologisch en Demografisch Instituut, Universiteit van Amsterdam, Amsterdam

Henk Voogd
University of Groningen
Faculty of Spatial Sciences
P.O. Box 800
9700 AV Groningen
The Netherlands

3 INTEGRATING ENVIRONMENTAL ASSESSMENT WITH DEVELOPMENT PLANNING

Nathaniel Lichfield

3.1 Focus

This paper continues the theme in my earlier paper to the 1992 Workshop in the University of Umeå, entitled Community Impact Analysis in Plan and Project Evaluation (Khakee and Eckerberg 1993). In that paper I started by summarising from past work the nature of Community Impact Analysis and ended by showing how it could be used for integrating environmental and planning considerations in reaching a planning decision on a project. This was exemplified by describing an actual case for office development in the London Green Belt, in Hillingdon near London Heathrow.

In this paper I pursue the same theme, applied not to a development project but instead to land use plans, policies and programmes as a whole. This application can be termed strategic as opposed to project evaluation.

I first show how the need for environmental assessment of plans, policies and programmes has evolved from that related to projects, starting with the United States National Environment Protection Act (NEPA) of 1969, which introduced environmental impact assessment. While progress in this direction seems slow, and the practice is far from well established, there is now a well recognised approach to Strategic Environmental Assessment (SEA) which is maturing. An introduction to the methodology is then given. From this base I then go on to introduce the role of evaluation in the planning process as a whole, as opposed to merely the project. These two strands are then brought together by showing how Strategic Environment Assessment and Strategic Plan Evaluation can be integrated. The indications are tentative when compared with those presented in the 1992 paper, simply because both environmental assessment and plan evaluation at the strategic level are still in the process of evolution.

D. Borri et al. (eds.), Evaluating Theory-Practice and Urban-Rural Interplay in Planning, 35–43.

3.2 Evolution of impact assessment on projects to plans, policies and programmes

The need for this evolution was seen in NEPA 1969 where it was visualised that EIA would be used not only for federal projects but also in their "land use plans, reasonable development programmes and economic programmes, including for those specific sectors" (Tomlinson 1986). This concept was further articulated in the Council of Environmental Quality guidelines of 1978, which interpret the requirements of the NEPA, that EIA may be necessary for the following actions (Therivel 1992)

1. "Adoption of official policy, set of rules, regulations and interpretations".
2. "Adoption of formal plans, such as official documents ... which guide or prescribe alternative uses of federal resources".
3. "Adoption of programmes, such as a group of concerted actions to implement a specific policy or plan".
4. "Approval of specific projects".

Similarly, the EEC also visualised from the outset of its environmental programme, in its preliminary draft directive of 1978, that the assessment system would apply to plans as well as to projects (Wood 1988). This proposal was not however incorporated in the Directive of 1985 which set up the environmental assessment of projects. But the intentions were clarified and emphasised in the Commission's Fourth Action Programme on the Environment in 1987 (CEC 1987):

> "the Commission's concern will also be extended, as rapidly as possible, to cover policies and policy statements, plans and their implementation, procedures, programmes ... as well as individual projects ...".

And it was endorsed in the Fifth Action Programme on the environment, agreed in 1992, as a necessary adjunct of sustainable development, namely (CEC 1992):

> "given the goal of achieving sustainable development it seems only logical, if not essential, to apply an assessment of the environmental implications of all relevant policies, plans and programmes....".

Pressures in this direction have also come from other international agencies such as the World Bank, the Asian Bank and OECD (Lee and Walsh 1992). The approach has also been officially adopted by the Department of the Environment in Britain, despite British resistance to a Directive by the EEC on this topic, on the grounds that it could be incorporated readily into the British planning system. Initially the policy was incorporated in the White Paper on the environment of 1990 (SSE 1990) which commmitted the Government to publishing guidance on the topic:

> "there is scope for a more systematic approach within Government for the appraisal of the environmental costs and benefits before decisions are taken.

> The Government has therefore set work in hand to produce guidelines for the policy appraisal where there are significant implications for the environment...".

The resulting guidelines presented by the DoE in 1991, and then again more specifically in relation to land use planning in 1992 (DoE 1991), again linked to sustainable development:

> "the planning system, and the preparation of development plans in particular, can contribute to the objectives of ensuring that development and growth are sustainable. The sum total of decisions in the planning field, as elsewhere, should not deny future generations the best of today's environment. This should be expressed through the policies adopted in development planning"....

Under these authoritative pressures, the practice of what has come to be called "strategic environmental assessments" has been slowly evolving in its theory, principles and practice From a recent review (Lee 1992) it is apparent that SEA has been widely applied around the world to a variety of strategic situations.

3.3 Methodology of SEA

That the extension of the environmental assessment beyond projects is desirable there can be no doubt. For one thing it would avoid the criticism of "*projectitis*" as opposed to more comprehensive planning. For another, it fits naturally as a stream of "environmental planning" into the development planning process in which are employed a number of varied sectoral streams, such as planning for traffic, open space, education. Thus employed, it would ease the preparation of project assessments and their approval, since these could be carried out within the framework of a general environmental assessment, just as development projects are in relation to a plan.

However, as indicated above, there is still no commonly agreed method for carrying out an SEA. This is perhaps not surprising given the widely varying circumstances within which the SEA needs to be applied, as brought in Lee and Wood (1978).

As the authors of the diagram indicate, the categorisation of action and type of assessment in the diagram "is a simplified representation of what, in reality, could be a more complex set of relationships". From this, it is understandable just why the SEA method cannot be standardised but must be capable of adaptation to both the particular purpose for which the SEA is being carried out (shown in the columns) and the level of government which is seeking the assessment (in the initial column).

Considering this evolution of EIA and SEA from their beginnings, it seems apparent that EIA can be regarded as a generic method of environmental assessment which then needs to be adapted for the SEA in the particular situation which decides the category of actions and type of assessment. Given this, it is to be expected that there would be similarities and differences between the EIA and SEA processes. This situation has been explored by Lee and Walsh (1992).

3.4 Role of evaluation in the planning process

Nature of evaluation

Evaluation is not a discrete but an integral part of the planning process. However, this is a comparatively recent innovation. Thus it was possible to say in 1975 that (Lichfield 1975):

> "Evaluation is frequently treated in practice as a discrete activity, functionally separate from other plan making activities, with those responsible for undertaking the evaluation work having little or no influence over the nature of preceding work. Alternatively, evaluation is left until too late in the study for it to make an effective contribution to subsequent decision making procedures. Awide variety of difficulties may result..."

In order to overcome the difficulties, and to enable evaluation to be better integrated in the planning process, the research used case studies in evaluation by the main methods which were then currently practised, namely threshold analysis and planning balance sheet analysis in regional studies; goals achievement matrix in sub-regional and urban studies; linear programming for a new town; and social cost benefit analysis for a major airport (Lichfield 1975, part II).

These studies showed that the particular planning process itself influenced the method, and vice versa. This has also affected our treatment here of the role of evaluation in the planning process: it has been necessary to adapt the evaluation process to the planning process within which it is to take place. Clearly there can be a bolder approach to evaluation in models which are not constrained, for example, by their relating only to the substance of physical planning; or evolving only the instruments of regulatory planning; or coming into the category of "market oriented planning".

For our purpose we have selected the development planning practice of Britain, post World War II. This label needs interpretation having regard to the changes since its inception in the late forties. Its essence can be grasped from the following:

(i) having the form and content associated with physical planning, with which is identified the social and economic activity implicit in the physical fabric;
(ii) operated through vertical integration in the national, regional and local levels, including budgeting and policy;
(iii) having the instruments of both regulatory and allocative planning;
(iv) having its plan making in the comprehensive rational, analytical mode, with implementation decisions being taken in a mode which is nearer "disjointed incrementalism" against the background of the plan.

But even within the one evaluation method and the one planning system there is no single application, just as there would not be for other inputs to the planning process, for example in survey and analysis. This was found to be so from practice where, in a large array of studies in Britain in planning balance sheet analysis/community impact evaluation, variations in the basic methods were adapted to reflect variables such as (Lichfield 1988):

(1) plans for projects or areas;
(2) kind of area plan (strategy or policy or comprehensive);
(3) scale of plan (from the national to the local);
(4) content of the plan (new town, renewal, conservation, roads, public transport);
(5) different aspects of a particular kind of content (private motor car, buses or pedestrian crossing);
(6) time and money resources available for the study (in depth to rapid assessment).

It was just because this variation between the case studies denied the use of a standard method that it has been necessary to formulate a generic method which could then be adapted as necessary to the variables found in the particular circumstances in any particular planning situation in any particular country and indeed in any particular state or province in a Federal country. It also can be adapted to the plan making, implementation and review phases of the British style of development planning process (Lichfield 1993). Evaluation is relevant in each phase, albeit for different purposes and to answer different questions. And since the three phases are themselves interrelated in a cyclical sequence, with feedback from the later to the earlier, the respective evaluation analyses are also so linked.

Sub-plans in plan making

The typical urban or regional plan is prepared for a specific area (the plan area) which is defined by a commissioning body. Since the plan area is often defined administratively, it is necessary in the plan making process to carry out studies in wider or narrower study areas which are functionally related to the topics in question, and thus not necessarily coincident with the plan area.
In content, the plan will typically show proposed changes within the content of areas or stability or "no planned change". For this purpose, there will be within any plan a number of interrelated "sub-plans" which typically cover such matters as the following, which we now describe in turn:

(i) policies;
(ii) proposals;
(iii) projects;
(iv) programmes of projects;
(v) strategies.

(i) Policies: A policy is a statement of intent, for the whole or part of a plan area, which will guide the making of decisions that arise on interpreting or implementing the plan. The conditions in which the policy is used could be formal (e.g. development control) or less formal (e.g. planning briefs to project planners within the area).
The use of policy as such comes within the field of "policy science", which aims to improve the quality of public policy making. As seen there, policy and plan making have

strong affinities to the use of the rational method. For example evaluation in policy making looks for inputs and outputs which are measured as far as practicable, as a basis for the evaluation.
(ii) Proposals: In contrast to a policy, a proposal relates to a site or area, as for example in use zoning. Thus while more specific than a policy it is less specific than a project.
(iii) Projects: Private and public sector schemes for specific sites, which carry with them or imply the allocation of resources for the transformation of the site within a visualised period, by means of specific agencies using specific implementation means (how, when, who, etc.).
(iv) Programmes of Projects: A number of projects throughout the area which are necessarily interlinked in space and time, since they have been visualised in the same planning framework;
(v) Strategies: Compared with the other elements, the meaning of strategy is more elusive. One appealing possibility is the military analogy of strategy and tactics. But while there is a common approach to forward looking in different fields, the field itself requires adaptation of the approach. So it is here. The strategy must be devised having regard to features such as:

(a) ends which are fundamentally socio/economic and physical;
(b) a mixture of developing and operating agencies, under various degrees of control by the planning authority;
(c) a long time scale for implementation of the "battle", which therefore implies programming for the short term and middle term;

Against this background strategy is the art and science of employing resources as a means (policies, proposals, projects and programmes of projects) towards achieving the ends implicit in the plan, in the recognition that those ends are in themselves the means in the chain for achieving the socio-economic values in the plan.

Adaptation of evaluation methods to sub-plan

Having reviewed the different kinds of sub-plan we now explore how the generic method of community impact evaluation for projects can be adapted to each. It will be relatively straightforward for "programmes of projects" (iv), since these can be seen as one major project with interrelated sub-divisions in space and time.
But the adaptation will be more difficult in respect of, say, regulatory proposals (ii), for while the input will be apparent it will be very difficult to predict the output of impacts with any precision in terms of development or timing. In passing it might be added that while this is clearly so in the British "flexible/discretionary" situation, it will be less so in a plan in Western Europe, where it is the plan itself as designed which is to be implemented (Davies et al. 1990). And the adaptation will also be difficult for policies (i), for these are merely statements of intention of the manner in which the planning decisions are to be made, in particular in relation to development control. Given the generalised nature of policies, and their proclivity for being in conflict with each other, there can be little certainly on the outcome. This situation can be overcome to a degree if scenarios are predicted as a guide to the intent behind the policies, for then it is the

scenario which can be evaluated on the assumption that it will be realised (Khakee 1991). For the same reasons scenarios will also help with evaluation of strategies (v).

3.5 Strategic environmental assessment and strategic plan evaluation

The approach

Having presented the integration of Environmental Assessment and Community Impact Evaluation for projects we now explore such integration at the strategic level, for policies, plans and programmes. The aim is to show how the generic model of project evaluation can be adapted to plan making.

The approach to this integration is well articulated in the Department of the Environment's Planning Policy Guidance on the topic (DoE 1992):

> "Most policies and proposals in all types of plan will have environmental implications, which should be appraised as part of the plan preparation process. Such an environmental appraisal is the process of identifying, quantifying, weighing up and reporting on the environmental and other costs and benefits of the measures which are proposed. All the implications of the options should be analysed, including financial, social and environmental effects. A systematic appraisal ensures that the objectives of a policy are clearly laid out, and the trade-offs between options identified and assessed. Those who later interpret, implement and build on the policy will then have a clear record showing how the decision was made, in the case of development plans, this should be set out in the explanatory memorandum or reasoned justification. But the requirement to 'have regard' does not require a full environmental impact statement of the sort needed for projects likely to have serious environmental effects".

This approach, it will be seen, suggests that it can be tackled by the adoption and adaptation of the generic model for the integration of Environmental Assessment and Plan Evaluation on projects. But clearly the strategic analysis cannot be as precise and quantitative as in project evaluation, bearing in mind that even here there is considerable qualitative judgement. Indeed the analysis of the effects must be general, indicative and non-site specific and the evaluation between options must necessarily be strictly ordinal (Verheem 1992).

But however qualitative the data, analysis and judgement, the aim would be a comprehensive evaluation of options, with a view to choice in the process of plan making. As such, it goes beyond the approach to the strategic Environmental Assessment in the preparation of the revised Lancashire Structure Plan, which used an "impact matrix" to predict the implications of 164 individual policy statements in the already completed plan for some twelve environmental components (receptors) that might be affected, with an indication of the comparative orders of magnitude (Pinfield 1992).

Adaptation for the Sub-plans in Plan Making

As introduced above (Fig 1) it is hardly practicable to be specific about SEA methodology since it can be applied to such a variety of situations, involving both the level of the plan, and also whether the SEA is to be applied to policies, plans, programmes or projects. But some guidelines can be given to the differences between the strategic assessments by reference to the indication above that the plan is not be seen as something distinct from, for example, policies or programmes, but rather that any plan contains within the one document a variety of sub plans which make up policies, proposals, projects, programmes of projects and strategies.

For this reason, in making the integration of SEA and Land Use Plans it is necessary to distinguish these particular components within each plan, in order that the generic evaluation methodology can be applied to each in its own way. For example, the end state of policies or strategies which are to be evaluated can only be crystallised by the development of scenarios which attempt to describe what would occur in the future if the policy, proposal or strategy were implemented. Projects would not be treated in isolation as in the generic model but rather all projects in the plan would be seen as "programmes of projects" interrelated in time and space and analysed as such. This would overcome the objection mounted for example against the evaluation of particular road proposals which are part of a wider network, whose accumulative impacts should be considered at the strategic level when account could be taken, for example, of synergistic impacts (the total impacts exceeding the sum of individual impacts) and threshold/saturation impacts (where the environment may be resilient up to a certain level and then becomes rapidly degraded) (Therivel 1992, chapter 1).

References

Commission of the European Communities (1987) Fourth Action Programme on the Environment, Com, Brussels, (86) 485

Commission of the European Communities (1992) Towards Sustainability: Fifth Action Programme on the Environment, CEC, Brussels

Davies, L. et al. (1990) Development Control in Western Europe, HMSO, London

Department of the Environment (1991) Policy Appraisal and the Environment, HMSO, London

Department of the Environment (1992) Planning Policy Guidance Note 12, HMSO, London

Khakee, A. and Eckerberg, K. (eds) (1993) Process and Policy Evaluation in Structure Planning, Swedish Council for Building Research, Stockholm

Khakee, A. (1991) "Scenario constrution for urban planning", Omega 19, 459-469

Lee, N. and Walsh, F (1992) "Strategic environmental assessment: An overview", Project Appraisal, September

Lee, N. and Wood, C (1978) "EIA - A European perspective", Built Environment

Lee, N. (ed.) (1992) "Special Issue on strategic environmental assessment", Project Appraisal

Lichfield, N. (1988) Economics and Urban Conservation, Cambridge University Press, Cambridge

Lichfield, N. et al. (1975) Evaluation in the Planning Process, Pergamon Press, Oxford

Pinfield, G. (1992) "Strategic environmental assessment and land use planning", Project Appraisal 7 (3)

Secretary of State for the Environment, et al. (1990) This Common Inheritance: Britain's Environmental Strategy, Commd 1200, HMSO, London

Therivel, R. et al. (1992) Strategic Environmental Assessment, Earthscan Publications Limited, London

Tomlinson, P. (1986) "Environmental Assessment in the UK: Implementation if the EEC Directives", Town Planning Review 57 (4), 460.

Verheem, R. (1992) Project Appraisal 7(3)

Wood, C. (1988) "EIA in plan making", in Wathern P. (ed.) Environmental Impact Assessment, Unwin Hyman, London

Nathaniel Lichfield
Dalia and Nathaniel Lichfield Associates
13 Chalcot Gardens England's Lane
London NW3 4YB
United Kingdom

4 EVALUATION IN ENVIRONMENTAL CONSERVATION PLANNING

Luigi Fusco Girard

4.1 Introduction

It has already been pointed out that the welfare-state crisis made it essential to find a new public/private relationship in the framework of new endogenous development strategies, so as to cope with the conflicts among different subjects, interests, objectives and values.
This study concerns the conservation and management of environmental/cultural resources that must be considered in a development-oriented perspective.
Today's successful local economies are those that can variously benefit local identity, regional specificities, the particular image of the place, the quality and therefore the attractiveness of a site.
But it is necessary, above all, that achieving such benefits does not mean just raising real-estate values, which would otherwise conflict with the general interests of an area: the goal is to integrate valorization with economic, cultural and social dynamics. The problem is to start a virtuous circle among economics, ecology and justice. These three dimensions' relationships have been destroyed by quantitative development, but the idea of *sustainable development* constructs this kind of integration.
The foregoing simultaneously concerns public institutions, private enterprises and the community of a region. An emerging central issue is a capacity to plan and control valorization in its multiple, differing (economic/financial, cultural and social) implications.
This means, above all, evaluating resources correctly and, if possible, in economic terms. But economic evaluation has always to be integrated with multi-criteria evaluation, which is possible if suitable evaluation procedures are used to compare alternative choices of valorization and management.
Multi-criteria analysis should be integrated in the perspective of considering the numerous subjects that have an interest in decision-making, using more and more a suitable software to support public institutions' decision processes. If this is done, it would contribute to reducing conflicts among interests, goals and values - *intrinsic* in valorization and in development - by creating alternatives in addition to those *given* initially that thus result from a process of active participation by the different groups concerned.

D. Borri et al. (eds.), Evaluating Theory-Practice and Urban-Rural Interplay in Planning, 45–55.

4.2 Valorization and Market

The increasingly heavy exploitation of environmental and natural resources by different economic subjects (industrial enterprises, real estate promoters, construction companies, etc.) is a serious threat to the sustainability of development (Blowers, 1993).
Current awareness of this makes it possible to use environmental assets as part of proper developmental strategies that integrate economics with ecology and justice. There is, on the other hand, the risk that a merely real-estate *valorization* process could be applied to environmental assets so as to make them marketable objects in a commercial system of exchange. This will end by eroding - because of bad or over use - the very value of environmental assets and by benefitting a few (e.g. owners, real estate promoters, direct users), excluding many people.
The state of public debt, a deficit-ridden economy and the like makes public financial resources less and less available. Where this is so, recourse to private capital, and therefore to the market, seems extremely necessary. We can talk of a *'restored'* importance of the market.
Privatization of environmental resources might seem a compulsory way to conserve them, but does the restored importance of the market entail acknowledging the free play of real-estate rent? Undoubtedly, as a tool, the market can promote new activity, demand employment and generate wealth, but it can also destroy or damage resources and values.
A *valorization* of environmental resources (e.g. coastal areas) can be realized by the market in terms of mere real-estate surplus value, i.e. acknowledgement of many rents that concern only the owners or promoters and omit several other subjects as indirect, potential and future users: i.e. anyone who is not a direct user.
A privatization/appropriation valorization in these terms that excludes a use with other subjects would, in the medium-long term, threaten the resources in question rather than protect them.
On the other hand, were a restored market indifferent to the issue of environmental protection, economic recovery would be very weak. It is enough to remember that the market produces externalities, or costs to third parties in the form of pollution, congestion, overuse and deterioration, which are heavy burdens on the land.
The behaviour of enterprises concerned only with short-term costs/revenues has caused the present - and would especially cause a future - ecological and social deficit. No *'good economy'* and *'good society'* can be founded on this kind of behaviour.
Externalities express how an uncontrolled, free market is unable to exploit resources in a sustainable way. Because they entail waste, and damage in the form of gross transformations of the environment, etc., they threaten the conservation of the assets in question.
The idea of sustainable development tends to conflict rather than accord with a free market. As a matter of fact, a free market enhances exchange values that are only some of the (possible) economic values of any rare resource (Blowers, 1993; Fusco Girard, 1983), that cannot be reproduced. This at the expense of other values as social use-value and independent of use value, as well as complex social value (Fusco Girard, 1986).
The realization of *sustainable development* entails some constraints on the uncontrolled exploitation of resources and therefore limits private ownership, corporate management, the activities of real-estate promoters and so on.

In the medium-long term, a valorization made in terms of mere exchange value can impoverish a resource and erode its indipendent of use value, its quality and beauty, as well as its social and cultural value; in the long term, it can destroy it.
Real-estate surplus value, if it is not realized correctly, can in brief entail costs in the form of cultural and social minus value. In managing conflicts between particular and general interests that the restored importance of the market produces, care is needed, an ability that, in turn, depends on a suitable evaluation of every value at stake and an ability to match them later in a balanced way. This means that the ability to manage the conflict depends on the suitable evaluation of resources and on conservation/development alternatives; as well as on enterprise ability in contributing also to the *good life* of everybody.
But it also depends on a consensus around sustainable development as being the general interest of a region.

4.3 Environmental resources evaluation

Evaluation is required to improve the processes of communication and of making decisions. It provides particular information that is doubly useful: in co-determining choices because of their correlation with an intrinsic dimension of human nature, i.e. decision making; and in being linked to another equally intrinsic aspect of human nature, i.e. to communicate, to participate, to live in relationship with other people.
Evaluation connects to two kinds of rationality: *instrumental rationality* that correlates ends with means, and *communicative rationality*.
The need to improve the information content of environmental evaluation arises from both kinds of rationality and from the acknowledgement that environmental evaluation has too often not succeded in transforming values in action that seek to avoid irreversible damage to the environment. Any judgement of the attractiveness of a resource - i.e. its ability to satisfy needs, its relevance, its *value* - depends on how different social groups (undoubtedly influenced by institutional or cultural conditions) perceive its attributes.
When compared to the other values and interests, environmental values do not usually become *actions* : in any conflict with these other values and interests, they do not prevail because their information content is too often hazy, imprecise, slippery or ambiguous.
As a result, it is impossible to avoid more or less irreversible damage to environmental resources, e.g. as a consequence of localizing or transforming a production activity, as well as of merely speculative real-estate valorization.
To make environmental values *operational,* it is necessary to improve and/or increase the assessed value of their informational content, so that they cease to be mere statements split from practical activity but can be *assimilated* into public or private debates and decisions.
Environmental evaluation should furnish arguments that communicate the *reasons* for protecting and/or improving environmental values face to face certain costs, which in other words states that some ends (conservation, protection, etc.) have a specific *value*.
To cope with this communication problem in order to determine decision process with the stake holders, one of the best way is through economic evaluation.

Evaluation using a monetary scale focuses on the economic potential of a resource, on its capacity to produce flows of benefits even at an economic level, on its being useful. A damaged resource is also an economic loss because of the loss of some benefits.
The goal of environmental resource economic evaluation is to communicate the qualitative or intrinsic value of a resource by reference to its utilitarian or *instrumental* value.
But thus putting environmental and economic values together on the level of interest that heavily influences public and private choices, this opens a way for environmental evaluation to *enter* into the decision process, can determine the priority choices, localizations, etc (Fusco Girard, 1983).
The theory of total economic value of resources that is briefly discussed below, expresses this effort of using *lower* order (economic) values to protect *upper* order (cultural, environmental, ecological) values.
This aspect of evaluation is connected both to what has been defined above as the instrumental rationality in the means-ends relation and to the rationality of communication. It is possible to establish a process of communication and/or dialogue between ecologists and politicians, economic subjects and members of the general public, between contractors and workers (Van Gigch, Roswall and Lagerovist, 1993), through which it is possible to make a social construction of objectives or goals in a rational way.
Environmental resource values can become a heritage not only of an élite but of everybody: something real, not an abstraction, being not only *good* and *right*, but also *useful*.
In this light, evaluation contributes operational content to the common interest and gives in particular actual and operational content to the idea of sustainable regional development that involves not only economic values but also environmental, judicial and equitable values. In the absence of such an evaluation, environmental, ethical and distributive values could not compare with economic efficiency and utilitarian values: they would be present only as an abstract *common good* of a region.
Such an evaluation transforms the often hazy, imprecise idea and principles of sustainable development as a regional goal into an operational concept that furthers the general interests of a region.
When evaluation is convincing, it can start a process of non-bureaucratic participation between multiple social parties and public institutions, or a learning or educative process that enhances decision-making.
In such a process of participation, each social party can compare itself with others and express its values; in the process, it learns other values. Because values are added and/or replaced, and orders of priority are changed, a different, more general viewpoint is developed. In brief, the community rationally chooses economic, ecological and social goals.
The foregoing enables us to conclude that evaluation is two dimensional in being able to rationally link ends with means and critical (making individuals more aware of and more responsible for choosing strategic ends for a community, an area, a region and so on).
Evaluation, communication and participation are closely connected to each other. How would it be possible to evaluate planning or management proposals without referring to goals expressing the general interest? On the other hand, evaluation also compares different ways of coping with a given problem and identifies the most attractive overall

solution or project, or the one that best satisfies the general interest by not only choosing between *given* hypotheses but also by developing new ones.
Because evaluation identifies the alternative that benefits most people, we can affirm that it is essential in that it promotes the ethical dimension in development, which is intrinsic to the idea of sustainable development.

4.4 The different economic values of environmental/cultural resources

It is very important to remember the existence of several points of view from which it is possible to get the economic dimension of environmental/cultural resources.
For the (eventual) owner only market value and its increases are important.
The promoter looks to maximize the difference between market value and costs, that is to increase the producer's rent to the maximum.

Fig. 4.1 Trend of different values in the time.

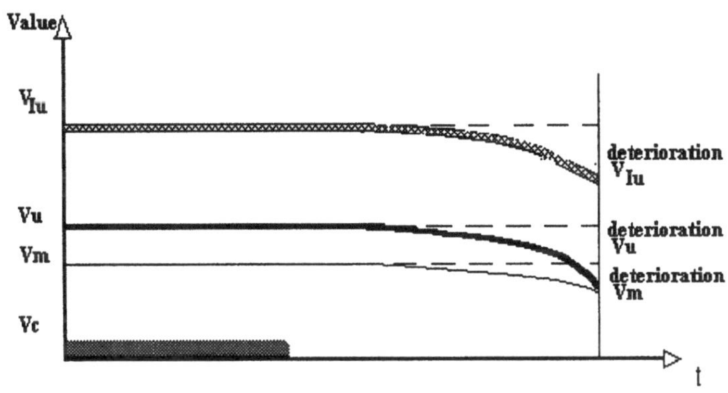

We must not overlook the point of view of the direct consumer, who is concerned with use value and also aims to maximize his income or consumer's rent, the difference between use and market values.
The user never considers market value, only use value and its maximization.
Finally, the point of view of the whole community represents the sum of social-use value for all kinds of users in time and space, and also for those willing to pay without having any use of the assets (independent of use value): we would say *total economic value* (VET).
Localization, non-reproducibility, rarity, uniqueness are important to understand the intensity of these different and co-existent values and their mutual combination.
In recent studies of the theory of total economic value, environment value has been expressed in monetary terms as follows (Pearce, Markandya, 1989):

Vet = Vuse + Vind use = Vuse direct user + Vuse potential user + Vuse future user + Vind use

In time the relations between these different values change, due to the resource context and to the activities aiming to contrast the deterioration of this resource.
In Figure 4.1 it is possible to see the trend of values over time and their possible combinations.
It is vital, in the process of privatization and/or valorization, to consider these different economic values.

4.5 Evaluation in the strategies for public/private conflict resolution

An economic surplus value, with a cultural surplus value and a social one might be obtained from a valorization of environmental resources that is consistent with the idea of sustainable development.
First of all, we should be able to identify the best relation among the different economic values we have considered: market, use, social use value, independent of use value. For example, if an alternative valorization improves a producer's rent while reducing that of a consumer, it will also reduce the social-use and independent of use values; it will entails a decrease in the asset's total value (see Figure 4.2).

Figure 4.2

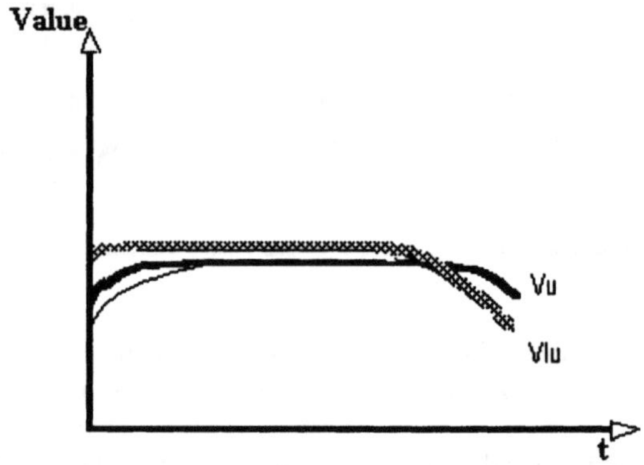

The opportunity cost of a valorization that appeals only to the market is the loss of that increasingly demanded value, through which it is possible to connote and identify a resource. This is the independent of use value.
On the other hand, if a valorization only increases use value and independent of use value and reduces producer's rent to its lowest level, it will have to rely only on public capital, which is less and less available at present.
So long as it is unavailable, this valorization will remain merely theory and not *action*.
An intermediate solution will be possible if the producer's rent rises only to a specific point, if necessary also by public charges carefully calibrated to activate a requalification/conservation process.

It is absolutely important to seek in turn the most suitable combination between public and private interests.
We have to find this public/private relationship so that both public bodies and private individuals may be *winners*: it means activating a positive sum game between conflicting interests.
Figure 4.3 shows the above. The solution A_1 achieved in terms of mere real estate valorization, allows maximum private income (for the producer), but a negative social utility. For instance, this happens when a heavy intervention is made in a given area, and benefits do not accrue to the area itself or the local community.
If public capital is lacking and no way exists to overcome the conflict so produced, solution A_2 is inapplicable because it would involve high opportunity costs for private interests.

Figure 4.3

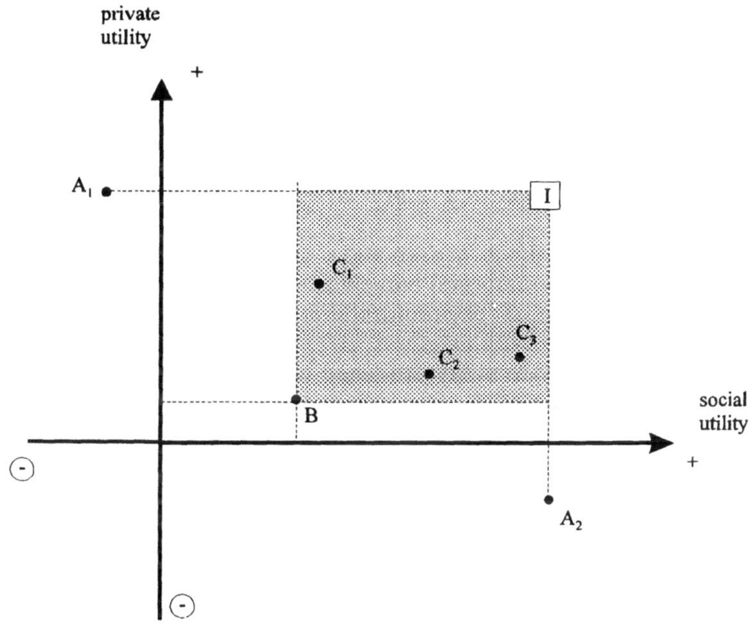

Solution I (ideal) would resolve the conflict between public and private interests intrinsically. This is not immediatly possible.
On the other hand, solution B shows the existence of a specific threshold for both public and private parties. These are minimum non-negotiable conditions (Keeney, 1992).
Therefore solutions lie in the shaded area in Figure 4.3
As defined alternatives become more numerous and possibilities grow of getting closer to solution I, the participation process gets stronger.
Evaluation of different valorization alternatives is essential, because this not only allows a comparison between *given* proposals, but especially stimulates the production of new

solutions. They will be more satisfying by reducing conflicts among interests, goals and values and by approaching the configuration I, where individuals realize that they - and all other individuals - may achieve maximum benefits. This would be the solution that comes closest to reflecting general interest.

4.6 Multicriteria evaluation in conflict solving

Multicriteria approach is necessary in analysis and in solving conflicts because it brings both tangible and intangible factors together. In fact, in conflict structure, economics elements are combined with extra-economic ones that include cultural, symbolic and perhaps also irrational elements. It is vitally important to consider these elements so as to break parties' inflexibility and to approach solution I.

Multicriteria/multigroups evaluation methods allow a more common reference model to be created to start with, that can include new (symbolic, cultural, etc.) dimensions through which to evaluate original hypotheses (Nijkamp, 1977; Nijkamp, 1980; Nijkamp and Rietweld, 1990). This means partial modifications that bring the best overall solution nearer.

Through the application of these methods, multiple parties (private individuals, firms, building promoters, owners, public subjects, etc.) can identify basic values and objectives and distinguish them from less important and marginal ones. These methods also activate a communicative dialogue in which a more advantageous general solution can be identified.

Multiple individuals disagree about conservation and development of environmental resources. Firstly, there are environmental movements and several public institutions that should guarantee the protection of environmental and/or landscape values; they aim to protect the public interest.

Promoters are interested in private utility produced by land transformations. Usually the achievement of the global utility of the first group occurs at the expense of the second group.

For instance, building promoters' monetary utility increases when areas at commercial/touristic destinations are extended inside and outside environmental assets. Then, the groups A and B have utility functions with inverse course, in the sense that losses for one are benefits for the other.

If the weight assigned to different functions is the same, equivalent convenience solutions for A and B will be those in which their losses are distributed (Mumpower, 1991) in an essentially homogeneous way (see figure 4.4).

Multicriteria/multigroups quantification-qualitative evaluation methods that can establish the complex value of every alternative, are useful in negociations between parties in dispute; they can generate acceptable solutions that overcome the logic of the zero sum game (Zeleny, 1992). Little by little, and enclosing other objectives/values, these methods add further objectives or values (by using both monetary and non-monetary scales) to extend the evaluation framework and to build winning strategies for every group involved: strategies in which improvements for both A and B are consistent.

There are several multicriteria evaluation methods to support public and private decisions that affect several individuals.

Some are more suitable for managing a lot of alternatives with very few criteria; others suit the opposite; yet others employ graphical descriptions to combine different criteria and visualize alternatives and priorities, which can simulate alternative environmental transformations. In other methods fuzzy approach has been introduced to define both objectives and constraints.

Figure 4.4

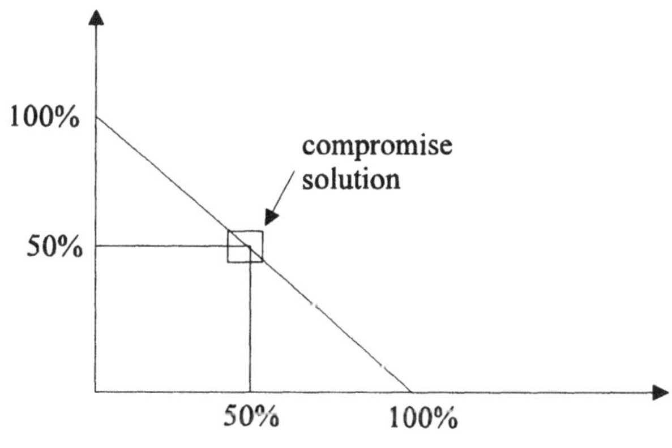

These evaluation methods differ from current ones because it is not just a subject who takes decisions, but a whole range of social and institutional subjects that interact, having their own (often conflicting) objectives, specific priorities and peculiar negotiable powers. For instance, utilization and valorization proposals for the same environmental assets might differ as between local, regional and national governments, labour world and entrepreneur sectors, etc. All this arouses conflicts that can seriously restrain or delay elaboration and implementation of conservation strategies.

A consistent application of evaluation methods can make it possible to mediate between different institutions and manage conflicts; in fact, a conflict can progressively be lowered to an acceptable level by working out new solutions. Consensus and cooperation would then replace it. These methods are thus a means of achieving an ethical dimension that is intrinsic to the idea of sustainable development.

4.7 Conclusions

Valorization of environmental resources in the light of ecological, economic and social sustainability implies careful and integrated evaluation that defines basic or strategic objectives in putting the idea of sustainable development of an area or a community into operative terms (Fusco Girard, 1993).

In this stage evaluation is an aspect of the communicative-dialogical rationality.

Evaluations have a central role in improving communicative processes and in incorporating the protection of qualitative values into the general objectives of a community.

Furthermore, evaluation is necessary to identify the choices that allow fixed goals to be effectively achieved. That is a technical dimension of evaluation.
Using economic-monetary and non-monetary evaluation procedures the integration of multicriteria/multigroups evaluation techniques is important, especially in the management-implementation, in order to resolve the conflicts between different subjects that are intrinsic to the conservation activity (Fusco Girard, 1992).
Multicriteria evaluation procedures should be introduced into the reorganization of the public sector to support choices and to make them more efficient, particularly in environmental sector.

References

Blowers, A. (eds.) (1993) Planning for a Sustainable Environment, Town Planning Association, London.

Fusco Girard, L. (1983) "Economic Theory and Valuation of Cultural Heritage", Restauro nn. 65/67.

Fusco Girard, L. (1986) "The Complex Social Value of Architectural Heritage", Icomos Information 1.

Fusco Girard, L. (1993) Risorse architettoniche e culturali: valutazioni e strategie di conservazione, Franco Angeli, Milano.

Keeney, R. L., Value Focused Thinking, London, 1992.

Mumpower, J. L., The Judgement Policies of Negotiators and the Structure of Negotiation Problems, in Management Science, vol. 37, 10, 1991.

Nijkamp P.(1980) Environmental Policy Analysis, Wiley, New York.

Nijkamp, P. Rietveld, P.(1990) H. Voogd, H. Multicriteria Evaluation in Physical Planning, North Holland, Amsterdam.

Nijkamp, P.(1977) Theory and Application of Environmental Economics, North Holland, Amsterdam.

Pearce, D. (1989) A. Markandya, E. Barbier, Blueprint for a Green Economy, Earthscan, London.

Van Gigch J., J. Roswall, B. Lagerovist (1993) "Setting a Strategic Framework for Conservation", Symposium on "Standards for Preservation and Rehabilitation", Fort Worth.

Zeleny, M. (1982) Multiple Criteria Decision Making, University of South Carolina Press, Columbia.

Luigi Fusco Girard
Dipartimento di Conservazione dei Beni Architettonici ed Ambientali
Via Cesare Battisti 15
80134 Napoli
Italy

5 EVALUATING SUSTAINABILITY: THREE PARADIGMS

Silvia Macchi
Enzo Scandurra

5.1 From unlimited growth to sustainable development

In 1972, H.D and L.D. Meadows published the book *The Limits to Growth*, a survey commissioned by the Club of Rome and carried out by the Dynamics Group of the prestigious Massachusetts Institute of Technology in Boston (MIT). This famous survey can be considered as having marked the official advent of the environmental question on a global scale, in that it placed the problem of the environmental limits to growth and the role of the market and of technological innovation in removing such limits at the centre of international debate.

The data collected for the survey were processed using dynamic models such as World-3, developed by John Forester of MIT, and belonging to the family of Doomsday Models. These models showed how in a world possessing finite resources, and small ability to find substitutes for these resources, the exponential growth of some sub-systems (population, agricultural and industrial production, pollution etc.) would prove to be critical factors in the inevitable extinction of mankind, caused by the unavoidable conflict between growth and the limits imposed on it by finite resources.

Even were productivity to increased, pollution to decrease, and the rate of population growth to fall, the predictions made by the model would remain substantially unchanged. The survey met widespread criticism from the scientific community, and particularly from many economists. While they accepted the survey's underlying theoretical assumption, that growth in a system of finite resources is necessarily limited, at the same time they held that this was unimportant. The problem of non-renewable resources could be solved by the market's powers of self-regulation, and by technological innovation's ability to find substitutes, they argued.

During the 1970s, economists' seemingly boundless faith in the market and technological innovation is apparent in the statements made by some leading authorities in the field. Beckerman, for example, claimed that as the increasing scarcity of all products tended to cause prices to rise, economic incentives would be introduced to encourage recycling and the use of substitutes. Moreover, he affirmed that man would not only find a substitute

D. Borri et al. (eds.), Evaluating Theory-Practice and Urban-Rural Interplay in Planning, 57–66.

for all resources that became scarce, but also raise the productivity of all types of energy and raw material as had been done at the time of the ancient Greeks (Beckerman, 1972).
Other economists, including Feige and Blau (1980), claimed that the application of neo-classical economic theory to the area of natural resources provided a powerful framework within which to analyze the problems arising and to consider the factors that would be important in determining the future exploitation of resources, which had always been notably undervalued in forecasting catastrophes. The most significant of these factors were certainly those relating to the "extraordinary possibilities of substitution" (in production and consumption) created by fluctuations in the prices of resources, and by often unpredictable and epoch-making technological innovation. Feige and Blau argued that these provided the alternatives that would satisfy human needs.
In reality, the general problems highlighted by the Forester-Meadows model, regarding the sustainability of development in a world of finite resources, have been replaced by the conviction that such absolute limits are meaningless. Today, the problem at the core of the environmental question is no longer the scarcity of natural resources (energy, raw materials, food, etc.), but rather the fragility of the environment, and the stability of the world's ecosystem. The Brundtland report, published in 1987, introduced the concept of sustainable development, or development that can satisfy our present needs, without compromising the ability of future generations to satisfy theirs - a major theoretical advance on the MIT report of 1972. The Brundtland report maintained that there are no precise population or resource-use limits to growth that, if exceeded, would precipitate ecological disaster. Different resources impose different limits on growth, and the over-consumption of energy, raw materials, water and land is manifested by rising costs and falling profits, rather than by their sudden disappearance (WCED, 1987).
Paradoxically, 'new limits' to growth emerge when breakthroughs in science and technology reinforce resource-substitution hypotheses, and when humankind shows it can 'invent' new resources as substitutes for traditional ones, which leads to an increase in the thermodynamic efficiency of economics and a concomitant reduction in pollution and waste.
Economic growth (that for many years has coincided with widespread affluence in society) implies a depletion of resources for future generations. The Brundtland report draws the attention of all the world's governments to the issue of sustainable development, or development that is capable of leaving to future generations the natural resources that current generations enjoy. The MIT report had drawn attention to the alarming process of the exhaustion of the earth's resources, and had warned of the danger of the extinction of the human race, if some kind of corrective models of growth were not introduced. The Brundtland report placed at the centre of the environmental debate the problem of the human impact on the planet and highlighted the medium-to-long-term consequences of this process.
The prevailing economic model has proved incapable of coping with the problem of finite resources, and, therefore, of evaluating the damage attributed by future generations to our depletion of natural resources that necessarily accompanies the process of growth. Yet herein lies a strange paradox: depletion of natural resources damages the earth and therefore future generations, while the business of repairing this damage increases employment, investment, levels of technology, and consequently Gross Domestic Product (GDP). Environmentalism seems to have become a major industry and welfare producer.

This social and economic paradox is apparent in every-day life: technological innovations in the protection of the environment (waste-water treatment and waste-disposal facilities etc.), the development of environmental engineering companies that evaluate the design of projects, all enter into the eco-business that increases society's wealth by contributing to GDP. In a sense, a country can delude itself that it has become richer, although it has really become poorer. Rather like an individual who believes he has become richer, because each year, in addition to his income, he spends capital (M. Bresso, 1993 b).
The consequence of this "dangerous asymmetry", as Repetto correctly observes, is that a country can exhaust its mineral reserves, raze its forests to the ground, erode its soil, contaminate its water reserves and hunt its animal and marine life to extinction, without the disappearance of these natural resources being registered in a drop in the country's income (Repetto R, et al., 1990).

5.2 The culture of evaluation

Facing up to the ecological emergency entails questioning some dominant paradigms in economics, in the so-called hard sciences (physics, mathematics, biology, etc.) and in the regional sciences.

The economic paradigm

Sustainable development requires the conservation of renewable resources for future generations, as well as those non-renewable ones (fossil fuels, minerals, etc.). The neo-classical economic model is unable to achieve these goals, either at a theoretical or a practical level. Neither the labour theory of value of Marx and Ricardo nor the theory of Pareto and marginal analysis is capable of attributing a value to natural resources, as according to the former, the production of such resources requires little or no work, and the latter sees natural resources as virtually infinite.
Economics textbooks tend to underplay the issue of the social and environmental costs arising from damage to the environment (Klassen G.A.J. Opschoor J.B., 1991). Neo-classical environmental economics, even in taking account of the interaction between the economy and the environment, considers pollution, for example, to be an externality, a public ill, resulting from production and consumption, that can be corrected by a number of specific regulations (Bresso M., 1993 b). It follows that protection of the environment is pursued through policies based on a principle of willingness to pay. With this approach, ecological equilibrium obtains when the level of pollution is constant, and when emissions do not exceed the capacity of the environment to absorb them. In practice, however, this equilibrium is all too easily disturbed.
In synthesis, neo-classical analysis postulates the substitution of natural capital (natural resources) by man-made capital, and believes in the ability of technological innovation to solve the problems arising from the scarcity of natural resources. According to its theory, therefore, the current model of economic growth can survive indefinitely, even in the face of scarce resources, provided these resources can continuously be replaced by man made substitutes, of being saved by technological innovation (resource saving), or, lastly, of being completely recycled (an impossibility).

After the 1970s, when ecology first appeared on the economic scene, the previously exclusive concern with operational techniques for resource conservation, gave way to the reconsideration of many theoretical issues.

Ecological economics has produced schools of thought each with its own theoretical approach: the energy balance-sheets of Ayres and Kneese (1969), the theories of the steady state and of zero-growth from Hardin (1968), Istock (1971), Meadows et al (1972), and Daly (1971), the co-evolutionary development of Perrings, and the thermodynamic development of Georgescu-Roegen.

These approaches converge in their interpretation of economic activity as predominantly the extraction of natural resources, transforming them via production processes into material goods, and eventually returning them to the environment in the form of effluents and waste. The originally closed circle of ecological systems is thus broken, when the manufacturing process transforms natural cycles into series of linear processes.

Indeed, according to Commoner, in nature the water, the carbon, the nitrogen and phosphorous cycles function as closed cycles. Nature knows no waste: the chemical substances extracted from the air, water, and the earth are recycled to form the raw materials for other natural cycles. Damage to the environment and pollution breaks these natural cycles: more materials are taken out than are put back, and waste-products increase at such a rate that they cannot all be naturally reabsorbed. The only way out of this quandary lies in urgent intervention by technicians, scientists and politicians to repair nature's broken cycles (Commoner B., 1971).

According to these schools of ecological economics, the production of effluents and waste-products cannot be considered an externality (as in neo-classical approaches), but on the contrary, an inevitable feature of the economic model. The thermodynamic approach of Georgescu-Roegen (1971 and 1976) developed an even more radical critique of the neo-classical model. The second law of his economic thermodynamics states that the use of raw materials and energy in economic activity entails an unavoidable increase in entropy; and at high levels of entropy, waste-products cannot be completely recycled, in the same way as it is impossible to obtain perpetual motion. Economic activity does not really imply any form of creation or destruction; materials and energy are first taken from and subsequently returned to the environment. What changes is not the quantity but merely the quality of items entering and leaving the economic system: between the extraction of the resources destined for production, and the release of waste-products into the environment, entropy increases, or energy is wasted. The inevitability of this increase in entropy implies that in the long term it is impossible to maintain a level of sustainable development, even in the conditions described by Daly's steady state that only prolongs the long term.

Overall, the attempt to found an economic programme on the definition of sustainable development provided by the Brundtland report implies a thorough-going revision of the theory and practice of economics, indeed of the very historically-evolved conceptual foundations of the discipline. Development is a co-evolutionary process, in as much as there exists continuous feedback between the environment and the economy, because in a growing economy, the environment is subject to continual change that in turn cannot fail to modify the laws of economics.

In this sense, the differing approaches of neo-classical, environmental and ecological economics may be capable of integration into a process of sustainable economic growth.

But even if they can, the present and future preferences of society with regard to the conservation of resources and the maintenance of consumption levels must form a constraint on the maximization of economic growth.

The paradigm of the natural sciences

Pursuit of the goal of sustainable development implies a reappraisal not only of economics but of the whole model of scientific reality on which the development of knowledge and technology has been founded, from Newton and Laplace onwards.
The inability of the market to address the environmental question, for example, is due not so much to problems of access to information, but rather to the fact that this information does not exist at the moment of decision-making, or when it is needed. The new concepts of unpredictability arising from Poincaré and Lorentz's intuitions and their subsequent formalizations offer little comfort concerning mankind's ability to predict and therefore to prepare for the future.
All decisions on environmental questions are taken in conditions of uncertainty. As to our initial state, we do not possess enough information about it nor can we ascertain the quality of the information we have. Nor can the mathematical models and simulations that are used explain a phenomenon or to predict its future behaviour be verified. Yet, in such cases, where 'facts' are usually uncertain, and their interpretation debatable, the stakes are high, and decisions must be made urgently. The result of such situations is that in many important environmental questions, hard strategic decisions are made on the basis of a soft scientific input (Funtowic S.O., Ravetz J.R., 1991).
Ecology invokes the laws of co-evolution: Perrings (1987 and 1991), O'Connor (1991), and Norgaard (1991) argue that the environmental and economic system evolved together. The use of natural resources and the discharge of waste-products deriving from economic processes cause continuous changes in the environment itself that in turn affect mankind through a complicated series of interactions. The ecosystem, the biosphere, has therefore continually to adapt to these changes and never achieves equilibrium. In these circumstances it becomes useless, even counter-productive, to try and control the expansion or destruction of a species, since the continuous changes imposed by co-evolution on the environment and on the economic system necessitate a constant revision of the conditions for economic and environmental stability.
We cannot maintain stability. The stability of ecosystems is tied to largely unknown variations in parameters and feedback processes. Scientific reason allows us to confirm that for some parameters (for example the proportion of carbon-dioxide in the atmosphere) we are far from the conditions of life of the biological ancestors of us as human beings. We cannot confirm with the same certainty however that the current trend towards increasing concentrations of carbon-dioxide emissions will lead to the extinction of the human race by a given date. In ecology, and more so than in other disciplines, our awareness that there is much we do not know and cannot predict apprises us of how infinitely cautiously we should act.
Our ignorance as to stability, or of our capacity to safeguard ourselves against our own disturbances of the environmental system (pollution, refuse, erosion, degradation, depletion of resources, etc.) should figure in any future evaluation that must therefore give up the use of rational, linear models, and use more circumspect methods.

The myths of rationality, indefinite growth and indefinitely available resources, the development of economics and the sciences, have created many problems from which we cannot easily escape. In this context, the question of "who decides?" becomes important. *What information is needed to take a given decision? Who possesses this information?* (These questions arise as soon as the interdisciplinary expertise that is certainly not easy to obtain is called for to tackle environmental problems). The very role of making decisions is thrown into crisis as soon as the consequences of earlier decisions must be coped with, while further changes can alter, or substantially modify, the premises on which the initial hypothesis was formulated. From this point of view, I believe that collective preference and the self-learning connected to the relationship between decision-maker and community that is described in the paper by John Forester, "*Ethics, Deliberation and Learning about Value in Planning Practice*", form a more fruitful approach to the complexity of the processes involved.

The paradigm of the regional sciences

The third paradigm is the regional one. The discipline of planning aims to regulate and control the complex cluster of activities that take place on earth, in cities and the countryside and are affected by human development. To reach these ends, planning has gradually developed technical knowledge that comprised increasingly sophisticated methods, instruments, and rules.

More than it has ever been, even recently, this knowledge is in crisis, and cannot effectively guarantee (and often not guarantee at all) the success of initiatives to control urban and regional development. Visible signs of its crisis can be observed in the actual experiences of local administrations that, despite disposing of a notable range of technology designed to carry out regional development, can rarely guarantee the success of an initiative proposed to this end.

The failure to reach the aims expressed by and contained in the various types of urban plans can be explained by the plans' intrinsically weak approach to the new (e.g. environmental) issues and their weakness for ever-more complex systems (such as the city) in which variables (population, needs, etc.) are unpredictable and ungovernable.

That these disciplinary tools (plans, regulations, etc.) still survive is due to their irreplaceable role that, by allocating land to particular uses, avoids the anarchy that would follow total deregulation. The entire power and expertise of town planning thus derive from its role as protector of those regions of the earth that ordinary human development has not yet jeopardized.

But this defensive role is not the only task of urban planning. In reality, every type of plan has dual functions: the one is to temper the process of transformation of the environment, through which the soil loses its function of reproducing resources) on one hand, and the other is to guarantee the presence of the essential conditions for an acceptable quality of urban life by providing green spaces, parking facilities and other services and achieving a means of urban growth that satisfies more cerebral considerations that include aesthetics, landscape and architecture.

Indeed, urban planning techniques have a role to play in resolving the conflicts that inevitably arise between a community's urge to acquire more and more land to satisfy its building and other requirements, and the equally sacrosanct wish to conserve those natural resources, sites, woods, forests, and ecological and biological systems that have not yet been jeopardized.

This allegedly unavoidable conflict of interest between human development and the environment clearly lies at the heart of debate. The purpose and power of planning seem to derive from this inevitability: as if, to satisfy human needs, human beings must unavoidably destroy nature while, at the same time, the constraints imposed by planning could really succeed in preserving the natural environment from any kind of human aggression. This idea, for example, has recently brought about an improvement in traditional planning tools, through the introduction of techniques to assess the future environment impact of a growth still considered compatible with the preservation of the environment.

Some disciplines that include economics and town planning are characterized by a property of operating on the real world, while others that include physics, mathematics and biology are characterized by a claim to knowledge. In those that operate on the real world, the current question is whether humankind can transform itself from a natural antagonist to the environmental metasystem into a producer of synergies with it. In other words, can a new alliance be forged between human beings and their needs, wishes, and creative instincts, and all that surrounds them and conditions their existence.

The problem that arises today, as we have already seen from economics and the social sciences, as well as from the hard sciences, is that of developing new categories of programmes and concepts, and of using new instruments with which to carry out these new methods of acting on the world. One way to get this process going is to attempt to define a number of terms and concepts that are too often incorrectly treated as synonymous. As we recalled above, the primary aim of urban planning is the optimal use of land; that of rural planning is to preserve the landscape; that of ecological planning is to avoid over-loading the carrying capacity of an area (Magoni, M., 1993). We can tentatively say therefore that environmental planning proposes to pursue these three objectives, and thus to move towards a peaceful co-evolution of the human and environmental systems; but this necessarily requires interdisciplinary expertise.

In urban and regional planning, the region is considered as merely an inert back-drop to human settlement activity. In environmental planning, on the other hand, it needs to be seen as a highly complex organism, a curious cultural and historical consequence of the systemic relations between the natural and human environments (Magnaghi, A., 1993).

The second step towards defining concepts, categories and methods is to define *regional sustainability*, at least as soon as it becomes the ultimate objective of environmental planning.

According to Magnaghi, *development* and *sustainability* are still being confused since the second is interpreted as a juxtaposition of remedial action and the limits to urbanization, but if we assume environmental damages are the results of exogenous and unsustainable patterns of development, the question of the sustainability of development needs to be put differently. It entails the transformation of the genetic rules of development itself, because the rules we have inherited from the model of unlimited growth can produce only unsustainable development (Magnaghi, A., 1993).

In this sense, Magnaghi directs our attention to the significance for sustainability of the regional approach. According to him, it is by definition both anti-economic in rejecting

sustainability that is subordinated to economic growth and anti-naturalistic in rejecting the environmental approach that attributes sustainability solely to natural ecosystems.
From these purely schematic concepts, it is possible to take several steps towards the *evaluation of regional sustainability*. At the planning stage of a settlement, we can, for example, refer to either the *diffuse* or the *compact* morphological model. But which can bring about an efficient use of a region (in the above sense)? We may say, for example, that the diffuse model implies more land (understood as a resource) and more private cars are used, while the compact model uses less land, and reduces the opportunities for travel, which offers an alternative solution to traffic congestion. Moreover, the compact model allows the optimal collection of refuse. Magnaghi's approach naturally assumes that local practice, the features of local sites, the existence of earlier settlements, and natural beauty and resources etc., all need to be borne in mind.
All these elements should be considered together, whenever regional sustainability calls for the presence of all social, cultural, economic and ecological opportunities, in as much as in the medium term they will determine the possibilities for development itself.

References

Beckerman W. (1972) "Economists, scientists, and environmental catastrophe", Oxf. Econ. Pap., 327-344

Berrini M., A. Campeol, F. Felloni and M. Magoni, eds.(1993) Aspetti Ecologici nella Pianificazione del Territorio, Grafo, Brescia.

Bottero M. (1991) "Ambiente, tecnologia, sviluppo: Un dibattito aperto", in Bottero M. (ed.) Spazio e Conoscenza nella Costruzione dell'Ambiente, F. Angeli, Milano, pp.81-101.

Bresso M. (1982) Pensiero Economico e Ambiente, Loescher, Torino.

Bresso M. (1991 a) "Riflessioni per la costruzione di un paradigma economico-ecologico", in OIKOS 3/91, 201-220.

Bresso M. (1991 b) "Il rapporto ambiente/sviluppo, ovvero come avviare la riconversione ecologica dell'economia", in S. Gentile, G. Zagato (eds.) La Riconversione Ecologica, F. Angeli, Milano.

Bresso M., Raffestin C. (1992) "Idee, valori e strumenti verso un sentiero di sviluppo sostenibile", in Sviluppo Sostenibile. Da un'idea Condivisa alla Progettazione Territoriale, International Seminar, IRRES, Perugia.

Bresso M. (1993 a) "Strumenti per lo sviluppo sostenibile: Questioni globali e locali", in XIV Conferenza Italiana di Scienze Regionali, AISRe, Bologna.

Bresso M. (1993 b) Per un'Economia Ecologica, NIS, Roma.

Common M., Perrings C. (1992) "Towards an ecological economics of sustainability", in Ecological Economics 6

Commoner B. (1971) The Closing Circle, Knopf, New York, USA

Costanza R. (1991) "Energia, incertezza, ecologia ed economia", in OIKOS 3/91, 167-200.

Daly H. (1977) Steady-State Economics: the Economics of Bio-physical Equilibrium and Moral Growth, Freeman, New York.

Daly H. (1991) "Economia ecologica e sviluppo sostenibile", in OIKOS 4/91, 97-118.

Daly H.E., Cobb J.B.jr. (1989) For the Common Good, Beacon Press, Boston.

Forester J.W. (1971) World Dynamics, Wright Allen Press, Cambridge, Massachusetts.

Georgescu-Roegen N. (1971) The Entropy Law and The Economic Process, Harvard University Press, Cambridge, Massachusetts.

Georgescu-Roegen N. (1976) Energy and Economic Myths, Pergamon Press, New York.

Klassen G.A.J., Opschoor J.B. (1991) "Economics of sustainability or the sustainability of economics: Different paradigms", in Ecological Economics 4 (2), 93-116.

Magnaghi A. (ed.) (1991) Il Territorio dell'Abitare, F. Angeli, Milano.

Magnaghi A., Paloscia R., eds (1992) Per una Trasformazione Ecologica degli Insediamenti, F.Angeli, Milano.

Magnaghi A. (1993) "Per uno sviluppo locale autosostenibile", in XIV Conferenza italiana di Scienze Regionali, AISRe, Bologna.

Magoni M. (1993) "Introduzione ", in Berrini M., Campeol A., Felloni F., Magoni M. (eds), cit.

Martinez-Alier J. (1987) Ecological Economics Energy, Environmental and Society, Basil Blackwell Ltd., Oxford.

Meadows H.D., Meadows L.D., Randers J., Beherens W.W.III (1972) The Limits to Growth, Universe Books, New York.

Meadows H.D., Meadows L.D., Randers J. (1993) Beyond the Limits, Chelsea Green Publishing Company, Post Mills, Vermont.

Pearce D., Markandya A., Barbier E. (1989) Blueprint for a Green Economy, Earthscan Publications, London.

Perrings C. (1987) Economy and Environment, Cambridge University Press.

Repetto R. et al. (1989) Wasting Assets: Natural Resources in the National Income Accounts, World Resources Institute, Washington DC.

Ruelle D. (1991) "Determinismo e predicibilità", in Le Scienze, 13-21.

Salvioni C., “Questione ambientale e sostenibilità dello sviluppo”, Working Paper, Università di Pescara

Scandurra E. (1992) “Verso una revisione dei contenuti disciplinari dell’urbanistica: il concetto di sviluppo sostenibile nella pianificazione del territorio”, in Sviluppo Sostenibile. Da un’idea Condivisa alla Progettazione Territoriale, International Seminar, IRRES, Perugia.

Schumacher Ernst F. (1973) Small is Beautiful: A Study of Economics as if People Mattered, Blond & Briggs, London.

Tiezzi E. (1974) Tempi storici, Tempi Biologici, Garzanti, Milano.

Tiezzi E. (1989) Il Capitombolo di Ulisse, Feltrinelli, Milano.

World Commission Environment and Development (1987) Our Common Future, Oxford University Press.

Silvia Macchi
Enzo Scandurra
Dipartimento di Architettura Tecnica e Tecnica Urbanistica
Facoltà di Ingegneria
Università "La Sapienza"
Via Eudossiana 18
00183 Roma
Italy

6 ECOLOGY, LANDSCAPE ECOLOGY, ENVIRONMENTAL EVALUATION AND PLANNING

Vittorio Ingegnoli

6.1 Introduction

An historical perspective is necessary if one is to focus on the path followed by the changes in the planning-environment relationship. We will see that the concepts pertaining to this relationship have changed at least three times in the present century, and we have now to work with the novelty of trying to join ecology and planning more directly in the concept of "sustainability" or sustainable development.
It is hard to say whether sustainability should be well defined as a concept or taken as an appropriate goal for management and planning. One way or the other, it calls attention to the management of systems under uncertainty, to the linkages between physical, biological and socio-economic systems, and to the interface between science and policy (Levin, 1993). Both ecologists and planners have to meditate about it because while its requirements can be reached in many ways, we do not know how scientifically and ethically true they may be.
For instance, we may consider that the ecological information about a territory is as important as other information - geological or socio-economic, or visual, etc. - about it, when all such information is linked in a mainly technical frame that refers to a mere geographical area. Or we may consider the ecological information as primary and linked diagnostically to all the rest, which implies a therapy: this is to consider a territorial unit as a living entity.
The complexity of these problems is great, while clarifying the right direction to follow is truly hard. There is no doubt but that we must change the methodologics of planning and many of its socio-economic goals, but must we also change ecology?
The question is crucial, because there are two contrasting possibilities. Ecology either defines a territorial unit as a) the geographical component of an ecosystem where human beings are foreign to a natural ecosystem, or b) as a specific level of life organization that forms a system of ecosystems, or a new living entity, with human beings as part of it. If either a) or b) is demonstrated to be correct, the reports of the discipline of planning would change sharply.
The consequences of these definitions clearly differ. For example, the well-known E.I.A. method of environmental evaluation would change: in the first case we could evaluate the

D. Borri et al. (eds.), Evaluating Theory-Practice and Urban-Rural Interplay in Planning, 67–78.

impact of human beings on nature, in the second (strictly speaking) we could not because no true conflict could occur, only an integration.

6.2 Environment and planning in Italy: a historical perspective

In Europe, environmentalism emerged in the second half of the 19th century in efforts to preserve monuments of architectural or historical importance, and natural sites of visual and recreational amenity. In 1852, for example, the city of Paris took over the popular Bois de Boulogne from the crown on condition that its woods and promenades would be cared for and improved. Similarly, a few years earlier, the city of London had authorized Victoria park. Later on, in 1884, the city of Milan planned the restoration of the Visconteo-Sforzesco Castle together with the design of the Simplon Park (Maniglio Calcagno, 1983).

After this early phase, we can identify in all the principal states of Europe, which were more or less anticipated by Britain three more phases that preceded the present.

In Italy, the first phase was from the beginning of the 20th century to the beginning of WW2 (1905-1940). Its dominant theme was the preservation of natural and man-made "monuments". After the first laws on ancient monuments (1906), and a large debate on both literature and architecture (Ricci, 1905) and nature (Serpieri, 1910), two laws were significant: on preserving woods to control geological risks (1923) and on conserving "beautiful" landscapes and amenities (1939).

The second phase (1940-1970) concerned the development of a framework of current planning (P.R.G., 1942 and 1967) and forest settlement. Its dominant theme was socio-economic control of the inhabited part of the territory, and a limitation on construction that was greatly enhanced after 1945.

The third phase (1970-1990) concerned an effort to develop land-use planning. The debate culminated with the law on landscape planning (1985) and the law on E.I.A. (1988). Its dominant theme was preserving scarce natural resources, underlining the visual and cultural values of the landscape, and requiring ecological information on remaining natural areas.

The start of the fourth phase has focused on sustainable development and is still weak. For the moment, the debate concerns mainly the concept of landscape in its geographical and aesthetic senses. Ecological information is important, but is equated with other, especially economical and sociological, information. Ecology, however, is more and more considered by planners.

But today we can find two cultural positions that seem to be far from an advanced ecology.

The first is common among those planners who confuse ecology with ecologism, often by mixing it with political sociology: participation is more important than biology, even when the environment is degraded. The second is common among those ecologists who consider man not to be part of the ecosystem and do not define the landscape as a system of ecosystems, with its own structures and functions.

For these reasons and in light of our introduction, it is truly necessary to try to expose some basic principles of advanced ecology.

6.3 Some principle of advanced ecology

As a discipline, ecology grew on biological roots to integrate organisms, the physical environment and human beings. Its capacity for such a complex integration is due essentially to these roots that permit it to define a "biological spectrum" (E.P.Odum, 1971, 1983, 1989). This spectrum shows a hierarchical concept of life, articulated in many recognizable levels of organisation (Table 1), which have in common the basic laws of biology, but different characters at different scales.

Table 1: Levels of Organizational Hierarchies in Biology

Physiological	Ecological
INDIVIDUAL	BIOSPHERE
Anatomical region	Biogeographic region
ORGAN SYSTEM	BIOME
ORGAN	**LANDSCAPE**
Tissue	Ecotissue
CELL	**ECOSYSTEM**
Citoplasm system	Biotic community
ORGANELLE	**POPULATION**
Biomolecule	*ORGANISM*

As shown in Table 1, the ecosystem level becomes the major focus at the ecological scale. Populations are considered as ecosystemic components and landscapes as associations of interacting ecosystems. Biotic community and ecotissue are intermediate levels that link population to ecosystem and ecosystem to landscape.

Populations, ecosystems and landscapes form the principal field of study of the discipline of general ecology, but each may also form a branch of it: population ecology that has been studied since the end of the 19th century, ecosystem or community ecology from the mid 20th century, and landscape ecology over the past few years.

It is important to emphasize that biologists and naturalists have studied the biological spectrum *in toto* only for 10-20 years. Its problem lies in the knowledge of the principal levels of life organization, and in some ambiguity of the concept of ecosystem (O'Neill et al.,1986). According to biotic or functional viewpoints, differences can appear between ecosystem groupings of components; moreover an ecosystem may be applied at any level of spatial scale. All that is confusing. We want to speak of advanced ecology, just to overcome the problem.

An ecosystem is a thermodynamically open system, far from equilibrium. Input and output environments are essential parts of it, which is as true for a forested area as for a city: what enters and leaves it as important as what is in it. This implies that there are, for each ecosystem unit, boundaries around it, a system of ecosystems outside them and population components within them.

Another basic implication of this concept is that environmental systems must be considered as biological systems: it is not possible to distinguish between a living organism and its environmental frame, for the organism interacts with and cannot live without its frame.

We are now able to answer the crucial question posed in the introduction: ecology (at least advanced ecology) defines a territorial unit as a specific level of life organization that forms a system of ecosystems, with a peculiar structure and dynamic. Human beings are part of a landscape, because their ecosystems are strictly integrated into the natural ones, as components of a system of ecosystems.

Today, we - human beings - are troubled by environment degradation, and especially that due to pollution; but on a wider spatio-temporal scale we have positively affected the development of the biosphere. Landscape bio-diversity has increased many times in a few millennia, and we may affirm that man has had an important ecological function in nature. This function is related to and expressed in many behavioural actions, firstly the tendency to plan an area and transform it and to maintain a territorial unit and its ecological stability.

Recently (BioScience, 1992), Eugene Odum drew up a list of basic ecological concepts and selected 15 on general ecology and 5 on human ecology. He enhanced it with a parasite-host model for man and the biosphere as a basis for turning away from exploiting the earth to taking care of it (going from dominion to stewardship, to use a biblical metaphor). In this sense, science indicates an urgent need to bridge the gaps between artificial and natural life-support goods and services and between unsustainable short-term and sustainable long-term management.

An advanced science of ecology that can study a mosaic of different integrated natural and human ecosystems represents a new challenge for the planning-environment relationship, because - we insist - the definition of landscape is a specific level of life organization. The consequences of this scientific view are very impressive: let us now check this point.

6.4 Consequences of ecological theory on environmental evaluation and planning

The consequences of this advanced ecological view are very impressive. In a wider philosophical sense it implies a change in the scientific paradigm.

If we admit the possibility (in fact the necessity) that different spatio-temporal scales lack linkages of simple continuity, we overcome the scientific reductionism (logical positivism). On the other hand, we observe that the subjective experience also changes (Lorenz, 1973, 1983), because it does not remain strictly limited to the private sphere of an individual.

Therefore, the new paradigm is centred on a multidimensional, multi-scalar concept of causality, in which the role of the observer has to be considered because it may influence results. This is true also for theoretical physics, in which the question was posed sixty years ago (Pauli, 1933-58), and then developed in many other scientific branches, as Risk Analysis.

But this new scientific paradigm becomes truly impressive in its consequences when it passes from the academic researcher's world to the field of advanced ecology and the concept of the environment. Therefore it implies changes in human behaviour and indicates new directions in territorial use and planning, in environmental evaluation and in socio-economical activities.

The fact is that the same concept of life organisation has changed: it does not remain limited to structure and functions of an individual organism, or their intrinsic components (e.g. cells). It must be enlarged to the entire ecological hierarchy, especially centred on the concept of ecosystem, and consequently on its components (populations) and its interacting formations (landscapes).

The definition of landscape as a specific level of life organization becomes a challenge for environmental planning and evaluation, firstly because man has to pass from a discipline related to technology, economy, sociology, urban design, visual perception, *and* ecology, to another related to biology, natural sciences, medicine, *and* traditional disciplines. Consequently, analysis, planning, and evaluation of the environment (e.g. territorial unit) have to change their methodologies: from engineering, economical and aesthetic rearrangement, to biological diagnosis and therapy.

In order to estimate the deep difference between traditional methods and the new ones, we want to check what happens to the concept of impact. The E.I.A. (evaluation impact assessment) was introduced 25 years ago in the U.S., then extended to all major countries. It is based on the concept of opposition between man and nature, therefore its goals are the minimization of impacts through check lists of action and components of design and an evaluation of their effects on the environment.

With some exceptions, this methodology contrasts with advanced ecology, particularly with landscape ecology, because here the concept of impact has no peculiar significance. The principle of integration between man and nature and the concept of incorporation of the perturbations in an ecosystem or in a landscape gets over the concept of impact (in Latin *impactus* means forced against).

We may speak of incorporation when the organization of an ecological system exerts control on some aspect of the environment that it could not have done at a lower level of organization. This permits: a) limits to alterations to its characteristics of stationary state, and b) use of perturbations as forces able to structure the system itself.

Thus, if a design is good and, step by step, follows the principles of ecology (and landscape ecology), it needs no E.I.A. This is especially true for planning.

As with medicine, the environmental evaluation needs comparisons with "normal" patterns of behaviour in a system of ecosystems. Therefore the main problem becomes how to know the normal state of an ecological system and/or the levels of alteration of that system. In medicine, the physiology/pathology relationship permits a clinical diagnosis of an individual, but the ecology/pathology relationship permits a clinical diagnosis of an ecosystem or a landscape.

Only possession of a good diagnosis allows you to devise a good therapy. In this case therapy includes a good environmental transformation, in the sense of compatibility *per se* and at the upper and lower levels of scale. As it is not possible to detail the applications of Landscape Ecology available for planning in this paper, we shall try only to rank some of them (Table 2).

Table 2: Some Principles of Landscape Ecology for Planning.

a. Analysis, referring to the *hierarchical levels* of spatio-temporal scale.

b. Definition of *a range of conditions* in which a biological system must remain.

c. Maintenance of the *mosaic patterns* and their compatibility.

d. *Congruence between structures* along the scales.

e. Management, referring to *thresholds of landscape evolution.*

f. Evaluation of the *disturbance incorporation* in landscape planning.

g. Maintenance of *boundary functions* (e.g. porosity).

h. *Moderate displacement* of resources.

i. *Appropriate connections* of the landscape elements.

j. Presence of *complementary patterns of stability* in landscape elements.

Table 2, which derives from previous work by the present author (Ingegnoli, 1993), shows many difficulties in order to study a territorial unit in accordance with the principles of an advanced ecology. For a planner, it is necessary to enter into a dialogue with an ecologist, in other words a natural scientist who has specialized in landscape ecology. However, any planner must have a basic knowledge of ecology. As expressed by Frank Golley (1989), landscape ecology frequently brings together information from a variety of other disciplines and interprets it in a form useful to planners, designers and managers.

In order to summarize how standard and ecological planning differ, we show in Fig.1 how their methods can be compared. Note that political and economical choices are placed in two opposite sequences, and that environmental impact assessment is not explicit in landscape planning. Anyway, the role of scientific information in political action remains the pivot of a new planning that aims at sustainable development. Moreover, ecological nonsense may result if ecological information is considered as no more important than standard planning information.

Figure1: Comparison between standard planning and landscape planning. Note the position (the role) of political and economic choices, and the difference among managing guidelines and landscape planning.

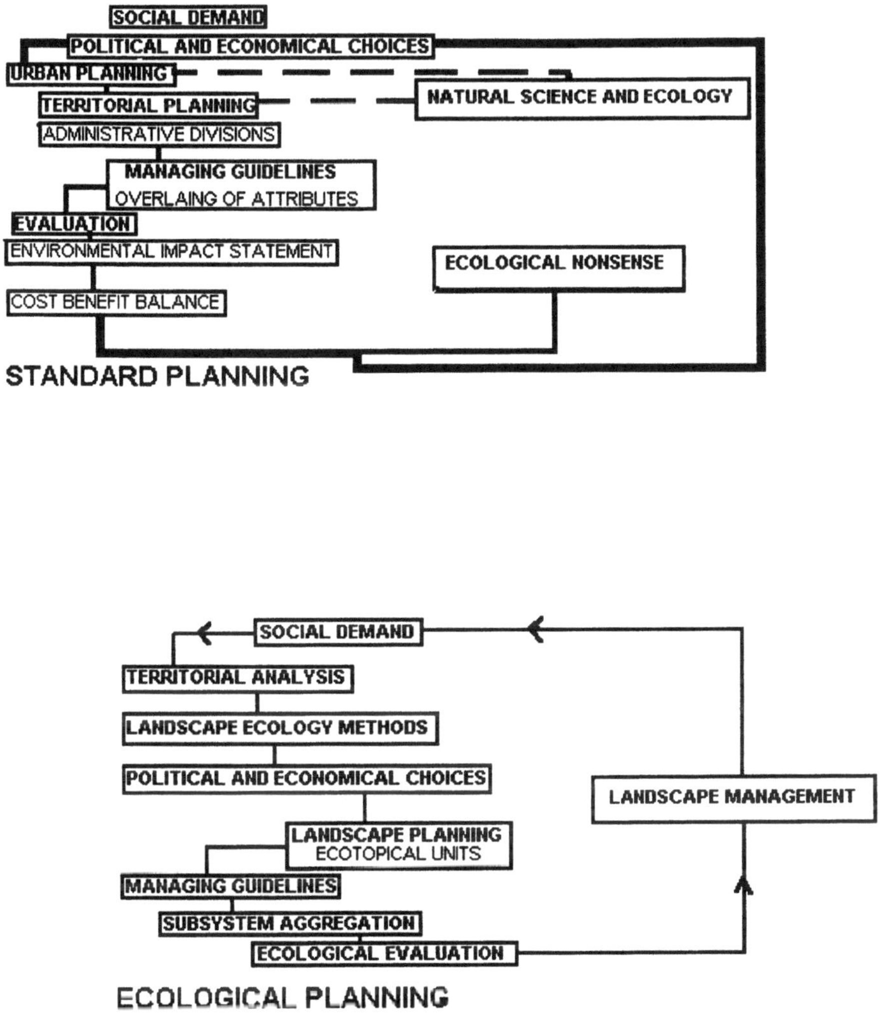

6.5 Obstacles to scientific information in planning processes

Wolfgang Haber, president of INTECOL, has recently indicated (Ecology International, 1992) seven major obstacles to the transfer of ecological research results into political action. If we add another problem, we may consider the main obstacles when the goal is planning and environmental management (Table 3):

Table 3: Obstacles hampering the transfer of ecological research results in environmental management.

1. Doomsday or chaos prophecy
2. The simplicity-complexity dilemma
3. Ecological fundamentalism
4. The value problem
5. The scale problem
6. Legislative maze
7. Dissent between ecologists
8. Different cultural education among planners.

Some of the reported obstacles must be emphasized, because they concern planning processes, especially those aiming at sustainable development.

The simplicity-complexity dilemma. Even ecologists find it difficult to explain the huge complexity of ecological systems, and to discuss it among themselves but nevertheless, we expect planners (and politicians!) and the general public to understand the ecological laws. Sometimes this complexity is too much for the human brain, which is why a mechanistic causality rather than a multidimensional web approach often prevails even among scientists.

The value problem. Even if problems are mainly ecological, and if we consider an advanced ecology that can integrate man with nature, the solutions are often not scientific but political, psycho-social, or, better, ethical. The value problem is moreover linked to ecological fundamentalism and prophecies of chaos (no comment!). Many politicians (and planners) refuse to accept the land ethic.

The scale problem. People rarely think on a time scale that exceeds three generations (grandfather to grandsons, or about 90-120 years), but no more than 40-50 years into the future. The political time-scale, obviously similar to that of planners, is shorter: 5-10 years, or less, sometimes as much as 20 years. The spatial scale is proportional to time scale. The minds of ecological scientists, by contrast, may move freely from the Pleistocene to the present and on to future centuries and, spatially, from the microsite to the biosphere. They are aware that a landscape has an average life of some centuries, and may receive influences from the level of a biome, to the home range of a small parasite.

Legislative maze. Old environmental acts and laws that remain in force confuse the new ones. The same field of application is entrusted to different agencies or institutions (e.g. landscape planning in Italy). Recent acts, as environmental impact assessment, are based on out-of-date principles and so the analyses they require are at least incomplete or unscientific.

Dissension among ecologists. All scientific disciplines have members who disagreed about basic concepts: science is driven by doubt, not by certainty. But ecology has a higher degree of dissent, because of its relative immaturity. Moreover, ecological planning seems to contradict the theory of self-organizing and self-maintenance of ecological systems, at least for scientists who are inspired by a reductionistic spirit. This is not true of landscape ecologists who still dispute fundamental problems.
Differing cultural education of planners. Planners derive from diverse academic disciplines: engineering, architecture, agronomy, forestry, economics and so on. They acquire differing cultural formations and are generally weak in the natural sciences (e.g. architects) or know only one sector well (e.g. foresters). Many are contaminated by ecologism, the environmental fundamentalism that contrasts with science; many others by political sociology or economics.
Ranking and emphasizing similar obstacles is necessary, because we have no other possibility than to know what is wrong if we are to overcome it: this may be considered as a first step. A second step is a propositive one and, when we speak of advanced ecology, especially landscape ecology, it is exactly in that sense.

6.6 Ecological applications: some new indexes, available also for planning

Many of the new ecological indexes in landscape ecology could be useful for planning applications: note, for instance, that the series of energy transformations in the hierarchies formed by self-organizing ecological systems are cascades of successive energy fractions; this explains why Mandelbrot's fractals often describe nature. Fractals, in fact, may be used as an ecological index (Turner 1989, Milne 1991).

$$\log(A) = d \log(P)$$

A = the area of a two-dimensional patch
P = the perimeter of th
e patch at a particular length-scale

d = the fractal dimension

When we speak of management referred to thresholds of landscape evolution, we should mention the BTC index (Ingegnoli 1991), a measure of biological territorial capacity of a system of ecosystems. The control of landscape changes is based on thresholds of metastability in human and natural habitats, in which BTC has a basic role.

$$BTC_i = 1/2\,(a_i+b_i)\,.\,R_i \qquad [\mathrm{Mcal/m^2/yr}]$$

where : $a_i = (R/PG)_i/(R/PG)_{max}$
$b_i = (dS/S)_{min}/(dS/S)_i$
R_i = the respiration of an ecosystem
PG = the gross production of an ecosystem
i = 1 to n = the principal ecosystems of the biosphere (or biome)
dS/S = R/B = the maintenance to structure relationship

This ecological index is also useful in order to evaluate the degradation of an ecotissue (or of an entire landscape), after an incorrect transformation. As we can see from figure 2, the variations of BTC in the Park of Monza in comparison with the regional and local BTC testify to the extent of alterations that have changed the function of that area in its landscape mosaic.

Figure 2: Variations of the biological territorial capacity (BTC) in the area of the Park of Monza (Lombardy), from its foundation to 1985 and with projections up to 2040. The ruled portion of the plot represents the *deficit* of transformation, and is to be compared with upper and lower landscape scale.

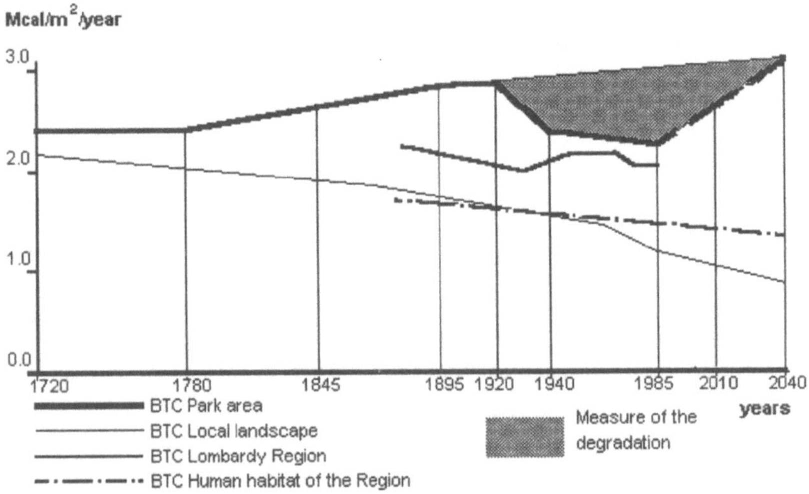

In conclusion, we should mention other new branches of ecology that have been founded to enhance evaluation and planning processes. Fundamentally, they are sectors of ecological applications, such as limnology and soil ecology. Pre-eminent interest is currently expressed in what is called system ecology, which concerns the interface between economics and ecology (Jansson, Zucchetto, 1985).

New ecological indexes, available also in economical applications, were presented by H.T.Odum (1988), and immediately adopted in system-ecological works. It is important to mention at least the concept of emergy (see the definition below). Odum perceived, from insights into the energetics of ecological food chains, that work should be redefined so as to distinguish kinds of emergy in the novel terms of it transformity (energy of one type required per unit of another).

Transformities may be used as an energy-scaling factor for the hierarchies of the universe including information. Solar transformities in the biosphere, expressed as solar em-joules per joule, range from 1 for solar radiation to trillions for categories of shared information. A joule [J=0.24cal] of mammal work and a J of sunlight are located in different parts of the energy transformation hierarchy, therefore it is incorrect to use energy as a measure of

work where more than one type of energy is concerned. We may express all resources in terms of the equivalent energy of one type required to replace them: this term is named emergy and may be defined as the energy of one type required in transformations to generate a flow or storage (in cascade flow of self-organizing ecological system).
Resource contributions multiplied by their transformities provide a scientifically-based value system for human service, environmental planning, public policies alternatives, and economic vitality.

References

Golley, F.B. (1989) "International dimensions of landscape ecology", Landscape Ecology 2/3, pp.137-138.

Haber, W. (1992) "On transfer of scientific information into political action", Ecology International 20, pp.3-13.

Ingegnoli, V. (1991) "Human influences in landscape change: thresholds of metastability", in O. Ravera (ed.), Terrestrial and aquatic ecosystems: perturbation and recovery, Ellis Horwood, Chichester, pp.303-309.

Ingegnoli, V. (1993) Fondamenti di Ecologia del Paesaggio: studio dei sistemi di ecosistemi, CittàStudi Editore, Milano, pp. 1-278.

Levin, S.A. (1992), "Science and sustainability", Ecological Applications, 3/4,pp. 545-546.

Lorenz, K, (1973) Die Ruckseite des Spiegels, Piper & Verlag, Munchen.

Lorenz, K. (1983) Der Abbau des Menschlichen, Piper & Verlag, Munchen.

Maniglio Calcagno, A. (1983) Architettura del Paesaggio, evoluzione storica, Calderini, Bologna, p. 316.

Milne, B.T. (1991) "Lessons from applying fractal models to landscape patterns", in M.G. Turner and R. H. Gardner (eds.), Quantitative Methods in Landscape Ecology, pp.199-238. Springer-Verlag, New York, pp. 199-238.

Odum, E.P. (1971) Fundamentals of Ecology, Saunders Company, Philadelphia, p. 574.

Odum, E.P. (1983) Basic Ecology, CBS College publishing.

Odum, E.P. (1989) Ecology, and our endangered life-support systems, Sinauer Associates, Sunderland, Massachussetts, p.283.

Odum, E.P. (1992) "Great ideas in ecology for the 1990s", BioScience, 42, 7, pp.542-545.

Odum, H.T. (1988) "Self-organization, transformity, and information" in Science 242, pp.1132-1139.

O'Neill, R.V., et al. (1986) A hierarchical concept of ecosystems, Princeton University Press, Princeton, New Jersey, p.253.

Pauli, W. (1933-58) Aufsatze und Vortage uber Physik und Erkenntnistheorie (On Physics and Knowledge), Franca Pauli, Zollikon, Zurik.

Turner, M.G. (1989) "Landscape Ecology: the effect of pattern on process", Annu. Rev. Ecol. Syst. 20, pp.171-197.

Zucchetto, J. and Jansson, A.M. (1985) Resources and society: a systems ecology study of the island of Gotland, Sweden, Springer-Verlag, New York, p.246.

Vittorio Ingegnoli
Dipartimento di Biologia
Cattedra di Ecologia del Paesaggio
Via Celoria 26
20133 Milano
Italy

PART II

PRACTICE

[Even when they want to,] "people
don't always say what they really mean."
UW 10/93

7 BEYOND DIALOGUE TO TRANSFORMATIVE LEARNING: HOW DELIBERATIVE RITUALS ENCOURAGE POLITICAL JUDGMENT IN COMMUNITY PLANNING PROCESSES

John Forester

7.1 Introduction

This paper explores the ways we learn about value in deliberative settings. In planning and many kinds of participatory processes more generally, such learning occurs not just through arguments, not just through the re-framing of ideas, not just through the critique of expert knowledge, but through transformations of relationships and responsibilities, of networks and competence, of collective memory and memberships.

The argument assumes neither that planning is participatory nor that it is democratic, just that deliberative conversations about value, about the interpretation and aptness of goals and means are inescapable aspects of practical action.

The paper argues in effect that many analyses of dialogue and democratic argument do not go nearly far enough to do justice to the learning that dialogical and argumentative processes can really promote. Inspired by liberal models of voice and empowerment, many analyses unwittingly reduce empowerment to 'being heard' and learning to considering seriously local as well as expert knowledge. In so doing, power is reduced to speaking and learning is reduced to knowing -- and the transformations of done-to into doers, spectators and victims into activists, fragmented groups into renewed bodies, old resignation into new beginnings, is lost from our view.

Much more is at stake in dialogic and argumentative processes than claims about what is or isn't true (as crucial and essential as factual analyses of health risks, for example, certainly are). At stake too are issues of political membership and identity, memory and hope, confidence and competence, appreciation and respect, acknowledgement and the ability to act together. The transformations at stake are those not only of knowledge, or in the last analysis, of class structure, but of people more or less able to act practically together to better their lives, people we might call citizens. This essay argues that participatory rituals provide participants not only with dialogue and argument, but with more of relevance than they anticipate, with more of value than they at first appreciate, with possible relations with others they could not foresee, and so with a literally

D. Borri et al. (eds.), Evaluating Theory-Practice and Urban-Rural Interplay in Planning, 81–103.

surprising, deliberative political rationality far richer than accounts of decision-making rationality or rational choice allow.

7.2 Two powerful models that help, but don't help enough: beyond understanding and dialogue to practical transformation

We learn from more than arguments and "voice" in participatory settings -- but how we do so is far from clear. In negotiations, participatory groups, and ordinary meetings too, we learn not just with our ears but with our eyes, not just with our heads but with our hearts. We come not only to hear new information that we find relevant, but we come to see new issues that need our attention. We come not only to revise our sense of strategies, but to develop new relationships with others too.

The public participation and participatory action research literatures draw often upon two powerful and related models of learning, both dialogical and argumentative, one indebted to John Dewey, Chris Argyris and Don Schön, the other to Paulo Freire and Jürgen Habermas.[1] Both are important, but both are too limited, needing to be refined and supplemented.

The Deweyan model focuses on the ways we learn in dialogical action together by testing our hunches and assumptions and suggestions of action. By making a move based on an initial strategy, being surprised at the consequences, whether positive or negative, and reframing our strategy as a result, we learn as we 'reflect in action' (Schön 1983). As a participatory action researcher working on community development issues, Ken Reardon, puts it,

> "Reflecting upon that information generates new assumptions, new theories, new hypotheses. But you don't just stop there: you take that knowledge and try to apply it to solve some immediate problems. In other words, you actively experiment with this knowledge you think you have acquired to see if it has the desired impact. It's this cycle that goes on."[2]

The Freirean model focuses on the ways we learn in dialogue by probing our political possibilities of speaking and acting together: who owns the land, controls knowledge, and who might yet have more control over their lives? Whose definitions of problems and solutions, of expertise and status, of power and powerlessness perpetuate relations of dependency and hopelessness?

The brilliance of the Deweyan model as refined by Chris Argyris and Don Schön lies in its reflective pragmatism, in its ability to make sense of the trial and error reflection-in-action of practical experience. The brilliance of the Freirean model lies in its insights into relationships of knowledge and power, voice and growth.

These insightful analyses have been very influential, but they are also inevitably partial and selective. They teach us a great deal, but they ignore a great deal too. Unless we supplement them both, preserving what they offer us but recognizing a good deal more, we will fail to realize the learning that participatory research (and deliberative democratic inquiry more generally) really can involve.[3] Without understanding the fuller promise of participatory processes, we will be likely to shape community meetings, workshops, retreats, mediated negotiations, and participatory research efforts in needlessly restrictive ways.[4]

This paper, accordingly, proposes a third approach that might be called, roughly, a transformative theory of social learning because it focuses not only on the ways that our arguments change in dialogues and negotiations, but on the ways that we change as well. At the heart of this transformative account is a view of participatory ritual performance in which actors not only pursue interests strategically and display themselves expressively, but reproduce and reconstitute social and political relationships with one another as well. Understanding participatory and deliberative encounters as social and political ritual performances means coming to see these as organized forms of presenting and exploring value rather than as going through meaningless motions, as forms thus connecting past memory and obligation to future strategy and possibility, and far more. Far from being empty containers in which dialogue takes place, these deliberative rituals are laboratories, if not cauldrons, of political judgment. But how does all this work?

In speaking together in more or less dialogical or argumentative settings, not only do we learn from what someone says or does about what they claim, but we learn still more from the way they do it. From the reasons they give, we learn about what others want or believe. But from the way they talk and act, from their style, we learn about who they are, "what they are like," what sort of "character" someone has (or is).

From the ways that others act, participants learn not only something about who these others are, whether they are arrogant or not, trustworthy or not, reliable or not, but how about they recognize, appreciate, and honor (or dishonor) values in the world we share with them. Reflecting upon an economically-focused analysis of a community market, Reardon's student team found it had misunderstood the entire institution. What mattered to the community members was not so much price but culture, not so much the economic marketplace as the social meetingplace. Action research participants learn about one another and about significant issues, too, from more than the deliberate arguments they hear. They learn, as we shall see in the next sections, not only from arguments but from incidental and quite revealing details, not only from the presentation of survey results, but from the surprises that their conversational rituals make possible.

7.3 The importance of messiness: letting the details surprise and teach us

In many participatory settings, of course, a good deal is unclear. Issues are formulated and reformulated. Relevant history is debated. Senses of future possibilities vary. Information is not perfect. Understanding is not guaranteed. Even if the words are clear, there's still far more to learn than meets the ear. A Norwegian planning consultant of much experience expressed this beautifully, in a simply and elegantly understated way, when he spoke of the difficulties of listening astutely in participatory processes: [Even when they want to,] "people don't always say what they really mean."[5] In that one line lie enormous implications for democratic practice, for critical listening, for the work required to appreciate and respect what people really do mean and value.

We learn not only from the points people make but from the details they present -- and often the unintended details. This is particularly important because many times we are at risk of being held hostage to our own presumptions of what others "really want." To respond to others and not just to our presumptions about them, we need to assess the particulars of what they say -- as bogus or accurate, as predictable or surprising, as genuine or exaggerated, and so on. Reardon speaks of the power of "additional information" and the process of considering it in changing participants' initial expectations:

> "No one goes in as a blank slate, with no notion of what they think they're doing there: what problems they're facing, what the independent variables are, and what the intermediate variables are and how they're going to impact those. But they get in there and they go through this very rigorous effort to collect a lot of additional information about these problems -- and all of a sudden it begins to change in its complexion."

In many meetings, the basic protocols, the basic ground-rules, can be pretty straightforward -- but the particulars of what is presented as significant are the sites of exploration, criticism, and learning. Each of the particulars, each of the objects of attention, has a place in a past and future story. Investigating the marketplace, Reardon's team discovered a rich history of relationships whose importance was not nostalgic but fundamental. Until the team understood the history of the market, they failed to understand what the market really was, what the market really meant to the people involved.

This leads us to stand the traditional fact-value hierarchy on its head. If value-free facts are, by definition, without value, really worthless, we can come to see that a claim about a "fact" is simultaneously a claim that something is important, that something matters. So too in participatory settings. The details presented are not mere details, worrisome minutae (though these exist to be sure), but they are often claims about value, claims about what one party is worried about, wants to gain, is afraid of, wishes to protect, or cares about enough to "put on the table" for discussion.

The particulars raised by others can seem irrelevant at first, and they may turn out to be irrelevant -- but they may also turn out to be surprising, suggesting problems or opportunities. Participants may come to see what seemed unimportant to be important, what seemed not feasible, to be feasible after all. This "coming to see" is a matter of recognition -- quite literally, re-cognition -- coming to see the very same thing in a new light, with new significance. In an initially health-focused project in Venezuela reported by Silverio Gonzalez and colleagues, for example, housing improvement turned out to be a surprising strategy that led in turn to new national health efforts to control parasites infesting certain kinds of campesino dwellings.[6] Seen now in a new political light, the particulars of housing improvements valued by the campesinos become not irrelevant but crucial to future possibilities of the government and international health agencies.

So participants learn not only about what others care about, but they themselves learn about concrete details and issues whose evolution or resolution will shape what they can do in the future, what they may hope for (or dread) today, who they can become tomorrow. In the Venezuelan case reported by Gonzalez, again, initial work on housing improvements led to further work and improvements on roads, water, and electricity for the campesinos.

But such learning from detail -- and unforeseen detail -- is hardly accidental. This learning is organized in a subtle way: not through a simple means to ends, but through varying rituals of meeting, talking, eating, and listening together. Reardon illustrates,

> "When we actually began sharing what we thought we were learning -- about how it would work and how it would need to change in order to be more like Wegmans -- we realized that wasn't at all the way people understood the importance of that institution. We were studying the wrong species of institution. And that's unbelievable when you think about it. We collected the surveys, we had all the statistics, we could report levels of significance. But it just didn't mean a goddam thing to the people we were working with." (emphasis added; jf)

In the sharing of concerns came surprise and learning, here, as well as in arguments about the truth or falsity of the survey's results. Promoting such surprise and learning, participatory rituals can enhance our own quite limited rationality too.

7.4 Letting stories supplement our limited rationality: reminding ourselves via ritual performance

PAR participants, like others in decision-making processes, are not all-knowing "rational economic men" with perfect information or perfect self-knowledge. Their predictions are uncertain. Their interests and senses of value, too, are ambiguous. In the flow of activity, they must focus their attention narrowly. But because they are not all-knowing, they can learn and be surprised.[7]

But like decision-makers more generally, PAR participants care about much more than they can focus upon at any given moment. They may be discussing job creation strategies or housing conditions, and so other concerns "not relevant at the moment" go unattended: the water quality of the neighborhood, the transportation system they depend upon to return home, longer run goals, and so on. But PAR processes may enable participants to learn not only from arguments about possibilities, but from allusions to what is relevant in the first place, especially when arguments, issues, alternatives and concerns are multiple, ambiguous, and conflicting.

Decision-making, planning, and participatory processes more generally are dances in which the initially relevant can come to appear irrelevant and the apparently irrelevant can come to appear relevant.

In planning and PAR processes, participants meet and discuss, propose and explore, even as they continue to learn about the implications of their own tentative offers and queries, their own contingent suggestions of "What if we...?" and "If we did this, could you do that?" and so on. Not going in with "a blank slate" as Reardon put it, not knowing as university researchers with Silverio Gonzalez what the campesinos could really do in Venezuela, PAR participants have to do a dance with their own intense directedness, their own singularity of focus, their own danger of concentrating so closely on yesterday's need and strategy that they miss today's and tomorrow's opportunity.

As participants realize the limits of their own knowledge, realize that they have to learn about relevance and significance, that all issues involved with practical options are hardly labelled ahead of time, they can begin to appreciate in new ways aspects of decision-making and participatory processes that appear at first to be needless preludes, ritualistic wastes of time, or even distracting preliminaries. Assessing an economic development case of participatory research in Norway, Morten Levin quotes a participant almost poignantly, "I know you have to force us to go through this ritualistic process of searching, but the results and the engagement that the conference ends up with are very exciting."[8]

Participants can come to see that ritualized small group meetings, for example, remind them of concerns they do have -- concerns with continuing relationships, needing to deal with each other next week, for example -- concerns they may have put out of mind in their urgency to deal with the issues at hand. They may see, similarly, then, that the ritualized turn-taking of story-telling reminds them of concerns they had forgotten: to take into account Garcia's new enthusiasm about working with Jackson, to be wary of Smith's easy promises, to watch out for the campesinos' being burned once again by campaign promises. They may see, too, in the turn-taking, what they "knew" but may have forgotten in their pressing desire to deal with the shelter's issues: that everyone present has their story, their sense of self, their varying political resources of support or opposition, of trust or suspicion, of cooperation or resistance (Viggiani 1991).

So PAR participants may see more generally that ritualized story-telling or small-group brainstorming or group identification of "strengths-weaknesses-opportunities-threats" functions in large part to re-mind them of their own concerns -- to bring into new focus values they have, obligations they wish to honor, interests they wish to satisfy even if they did not have all this in mind at the beginning of a given meeting.[9] Information begins to take on "a new complexion," Reardon says, and in related participatory action research involving the organization of an urban women's shelter, Michelle Fine tells us of the need to structure a continuing conversation in which issues can be explored, in which diverse stories can be heard.[10]

The official, recorded "minutes" of previous meetings often fail to accomplish this ritual re-minding process. The words on the page, typically quite abstracted from their context and summarized quite drastically, pale in comparison to participants' actual voice and style as they present themselves and reveal their concerns (or demands).

Importantly too, a simple summary list of issues may never do the job of the initial story-telling, precisely because the summary list is too simple. When issues are complex, when organizational decisions involve or will affect many actors, decision-makers (PAR participants) need reminders that will help them identify emergent issues that will matter, particular issues of this new dispute or opportunity and not just general topics of concern.[11] So the detailed richness of stories, their apparently distracting detail, can serve participants by reminding them of interests and teaching them about issues they may hardly have had in mind at the meeting's beginning.[12]

7.5 Learning about value in ritualized story-telling processes

Far more than descriptions of events, stories are forms in which their tellers can offer what they take to be worth passing on. Telling stories, participants present examples of virtue to emulate or vice to avoid: campesinos control of their own housing improvements to encourage or "outside-in" centralized development strategies to avoid. Telling their stories, participants present promises honored or betrayed, offer insights regarding the "real story" behind another's facade, or recount a history of "who did what to whom," for example, when the Governor's last program was implemented. Such story-telling provides not simply a "Who Dunnit?" record, but the quite practical context for future action and judgment (Forester 1993).

Because they present what speakers take to be worth telling in the circumstances of addressing the problems at hand, stories are a conventional form in which PAR participants articulate value. In the design of PAR processes, then, the deliberately ritualized organization of public story-telling, we see the ritualized means through which participants can articulate and explore value together. Of those practitioners we have cited above, Levin refers to these as organized "fora" for the exchange of views; Fine refers to this as furthering a needed conversation about tough issues; Reardon points to the importance even of "schmoozing and hanging around and informal discussion" to learn about what really matters to community members.

Such participatory rituals of telling and listening to stories can work transformatively in at least three ways: to transform identities, agendas, and perceptions of value in the world. Consider each briefly:

7.6 Transforming relationships and identities

First, and perhaps most commonly, conversational or story-telling rituals can produce or reproduce, strengthen or weaken, the public senses of self we call "social identities." So, for example, with the mutual acknowledgements of "Do you...?" and "I do," a marriage ceremony publicly transforms (the identities of) two "single" people into (the identity of) a "married couple." Similarly, a weekly religious ritual of participation in "church" can be embraced (or resisted) as it presents its participants, more or less authentically, as practicing members supposedly honoring a religious tradition.[13] So Reardon refers to participation in the community planning effort as a "developmental process" in which people come to "carry themselves differently," and to a central dimension of learning in participatory processes that we must not neglect:

> "[D]eveloping the capacity to provide leadership. . .is one of the critical criteria one would use in evaluating the effectiveness of action researchers. It's not just that they've come up with some interesting solutions to three or four immediate problems that you'd consider to determine whether the action researcher's done a good job. The question is also: what's the ability of current residents and leaders to continue the process and to train other people in their own community..?"

In participatory processes, the ritualized interactions can shape new identities and relationships. New groups, organizations, or networks, not just arguments, can be created. Participating together to promote economic development or to provide more responsive services, we form new working groups, sub-committees, task-forces, and networks. As a result of meeting together, making cautious and contingent agreements together, exchanging tentative promises of "I will do this, if you do that, and then we'll see what else we might do. . ." participants recreate and redefine their relationships, creating at times practical new organizational forms and networks.[14] "Know-what" and "know-how" come to be supplemented here with a particularly social "know-who."

7.7 Transforming issues and agendas

Second, rituals pattern attention selectively. Writing lucidly about the political significance of such ritually organized attention, Steven Lukes argues for a study of ritual:

> "[which] would go well beyond the conventional study of politics, which, as Edelman puts it, concentrates on who gets what, when and how, on "how people get the things they want through government," and focus instead on mechanisms through which politics "influences what they want, what they fear, what they regard as possible and even who they are" (Edelman 1964:20). . . . It would also examine the ways in which ritual symbolism can provide a source of creativity and improvisation, a counter-cultural and anti-structural force, engendering new social, cultural and political forms, involving what Turner calls "liminality" and "communitas"..."[15]

We can see rituals, in Lukes' view, then, "not as promoting value integration, but as crucial elements in the "mobilization of bias."" (Lukes 1975, p. 305).
In their ritualized forms of collecting stories, brainstorming, recording and reporting small group discussions, creating larger group memory by covering walls with lists of discussed items and topics, PAR participants necessarily focus their own attention selectively. Coming together, they invoke and honor certain values and not others; they read or quote certain texts and not others; they express and respond with certain emotions and not others.[16] They express as a matter of public record certain allegiances and not others. Providing as Lukes puts it, potentially "a source of creativity and improvisation, a counter-cultural and anti-structural force, engendering new social, cultural and political forms," these participatory and deliberative rituals prepare their participants -- more or less well -- to recognize new issues, to attend creatively and responsively to particular problems at hand.[17]

7.8 Transforming ends: what's at stake

Third, then, ritual performers articulate, organize, and invoke claims of value together.[18] In a search conference devoted to the initial exploration of strengths and weaknesses, threats and opportunities, the ritual participants are asked to step back from a "last word" analysis to share with one another their informed hunches about what matters, about what is worth considering together, positive and negative. Their ritualized stories may often be richer than anyone intends, and so the very occasioning of story-telling allows surprise and learning that extends beyond the narrow purpose of any single story-teller.

We say more than we intend, inevitably, and we mean even more practically than we say.[19] Perhaps the most brilliant expression of this point comes in a line of Iris Murdoch's, which might help us not only to listen to stories but to look carefully, too, at their sensitive and wise telling -- in participatory and deliberative conversations, planning and policy making conversations, and those of everyday life too. "Where virtue is concerned," Murdoch wrote, "we often apprehend more than we clearly understand and grow by looking." (Murdoch 1970, p 31).

Listening together, we recognize as important not only words but issues, details, relationships, and even people we may have ignored or not appreciated in the past. Listening to tone as well as content, we may recognize not just claims about what the government has or hasn't done in the past, but a history of betrayal and resulting fear, suspicion, distrust -- which must be acknowledged, respected, and addressed if working relationships are to be built. Recognizing such issues in public, acknowledging them in participatory groups, we come not only to transform our shared senses of what's at stake today in our deliberations but to shape new commitments to new ends and to one another as well. Our learning, our transformation, is both cognitive and collective: in acknowledging another's concern we can come to see it more clearly, testing "to see if we have it right," and we can encourage and shape relationships of solidarity and collaboration too.

But not all stories or rituals are created equal. We have all been there: Smith may drone on and on, not presenting issues of value but boring everyone to tears and driving them to distraction. Epstein's story may in fact be irrelevant or an absurd paranoid fantasy. Jones's story may be obnoxious or distracting or ill-timed, all of which may threaten and not serve a participatory decision-making process.

So not any turn-taking, sharing of stories, or generating of shared texts will do. If the participants do not have some shared sense of the "rules of the game" in their meeting together, they may not be able to act together. If they do not have some sense of structure and process, of protocols of turn-taking, of appropriate and inappropriate action (and story-telling!) in this meeting, they may be too confused or threatened or shy or reticent to participate, sharing what they know, signalling important concerns to others, reminding others, warning others, showing others new options too.[20] If participants do not feel safe and trust each other enough to speak and listen responsively, they will hardly work together, and their negotiations and participatory decision-making will most likely fail.[21]

7.9 The ritual structuring of unpredictability as the ground for learning, or decision-making when interests, parties and priorities are changing

But of course meetings can be so structured, so predictable, so pre-determined that no-one learns much of anything new. The designers of the decision-making process may have decided what information shall be explored and how it shall be, and their controlling intentions may make surprise, discovery, and the identification of new issues and opportunities practically impossible. Of course if someone could really know ahead of time just what issues and information were going to be relevant, he or she might not only design wholly predictable decision-making processes, but predict, if not dictate, the precise outcome as well.

But if, more realistically, we know that we do not know everything that will be relevant, if we know that we do not know what options we will discover in the process of listening and responding to one another, then we need not so much rigid predictability but instead a structured unpredictability that will help us to ask new questions and consider new answers. This structured unpredictability of ritualized story-telling may be the most important designable element that facilitates practical learning by participants about the breadth and depth of their own concerns: it exposes them to relevant but surprising, important but unforeseen, claims (facts and issues, provocations and emotional appeals, and more) that they take to matter.

This unpredictability of the ritual itself -- the search conference, for example, or the neighborhood planning process's review of data -- enables emergent groups to form: a steering committee, a sub-committee, an e-mail group, a network or coalition, and so on). The ritual occasion may structure the safe collaborative possibility of participants exploring and forming tentatively new roles, new groups, and thus new identities, along with accompanying newly designed norms, rules, agreements or conventions that articulate how the participants may go on together. The openness of a search conference or a mediation session to a new agreement about roles and rules means that ritual occasions enable innovative self-generation, group redefinition and innovation, that may not be envisioned by any participant at the meeting's onset. Recounting his team's work in Venezuela, Silverio Gonzalez provides an example,

> "Everybody (campesinos, regional government, university, health authorities) wanted that positive action [of effective housing improvement], but nobody believed it was possible. "Verbal agreements" are abundant in Venezuelan public policy, but when there were actions according to those agreements, [there came] an explosion of sense and energy among participants."

If surprise is essential in the deliberations of actors with limited information and rationality, participatory rituals are the social structures that may stage and enable the unpredictable learning of such surprise.

7.10 From garbage cans to transformative rituals

James March and his colleagues have described a somewhat analogous process of non-intentional discovery in decision-making settings. Painfully aware of participants' limited knowledge and time, shifting preferences and resources, March and his co-researchers chose a somewhat desperate metaphor as they characterized a "garbage can model of decision-making." In the garbage can, streams of choices, problems, solutions, and participants meet in unpredictable ways to shape on-going, complex and messy organizational outcomes. As March put it,

> "[O]rganizations. . . provide sets of procedures through which participants arrive at an interpretation of what they are doing and what they have done while in the process of doing it. From this point of view, an organization is a collection of choices looking for problems, issues and feelings looking for decision situations in which they might be aired, solutions looking for issues to which they might be the answer, and decision-makers looking for work."[22]

Understood in this way, participatory decision-making becomes a good deal richer than even an elaborate bargaining process.[23] Participation itself is assumed to be fluid, with some people paying more attention to some issues than others, having more time at some points than others, and so on. Participants clarify and discover goals rather than take them as given, and so in the process of interacting they come to value some ends they had not cared for previously. Similarly, techniques become not only tools to be used upon demand but competences whose presence shapes participants' search for solvable problems and understandings of themselves. Electronic mail capacity, for example, will shape participants' thinking about their needs to keep in touch with one another easily; "have computer, will network" the participatory action research saying might go, updating the more common, "Have hammer, look for nails..."

But the "garbage can" model of decision-making can itself be refined or recycled. Choices, problems, participants, and solutions may not so much be dumped into garbage cans, as March and his co-authors suggest, as they meet, interact and transform one another, in ordinarily structured ritual performances that organize members' attention and shape their emerging exploratory commitments.

Held hostage by its own metaphor, the garbage can model of decision-making and deliberation is right, but not right enough. Participants and problems and opportunities do often come together in ill-structured and messy ways, do often find a voice, or fail to, in unpredictable ways, but then more happens too. In their ritualized interactions, participants can come to see one another in new ways; problems can be redefined and reformulated; opportunities can be clarified; priorities can be reordered individually and collectively. The process of deliberation and participation, then, is better seen not only as argumentative or dialogical in terms of who knows what, not only as allocative, in terms of who gets what, but as transformative too, in terms of who comes to create new relationships and act on new commitments in actual practice.

Participants begin deliberative processes with agendas and suspicions, with complex cares, sets of interests, and senses of possibilities -- all informed by yesterday's, but not yet informed by today's, experience. By articulating and exploring their ambiguous priorities, evolving relationships, and uncertain options today, participants in ritualized processes of dialogue and deliberation can transform their expectations and obligations as they make new agreements or revise old ones; they can transform their senses of self and other as they take new roles or redefine old ones; they can transform their senses of value and priorities as they come to recognize new issues or re-evaluate old ones; they can transform their working relationships as they form new organizations and networks or reshape old ones. Far from finding themselves simply thrown together as chance may have it, as March's garbage can model might suggest, deliberating participants can have a good deal to say about what they do together, about how they listen to one another or fail to, about how they search for new options or fail to, about how they learn and find new ways of going on together, or fail to. They have a good deal to say, then, not just about how they can refine their knowledge of strategies but about how they can transform themselves too. Reardon spoke of this possibility in East St. Louis:

> "This type of support is really giving them power, empowering them, and exciting them, and it's paying off for the community. You can begin to see people change in the process: they are more likely to voice their concerns. If they are not listened to, they pursue the point and feel like they have a right to do so, whereas before they might not have. They have learned not to back down, to feel like they have a right to negotiate with people in power over important things and have a right to participate in the process. For many people, that is an incredible leap in their consciousness and sense of confidence in what they deserve and who they are. They carry themselves differently."

The process is one not simply of throwing together a group of rational decision-makers, but one of changing those decision-makers into a more deliberative political body. In this way, the ritual richness of participatory encounters provides the infrastructure, the materials and occasion, for a deliberative political rationality, in which means and ends and self and other are transformed, one far richer than typical optimizing accounts of rationality allow. But political power is always an issue here. So Gonzalez, for example, reported not only the campesinos' increased political organization, their successful struggles for roads, water, and electricity, but the resulting manipulation by the ruling party.[24] Reflecting upon her work with a women's shelter, Michelle Fine speaks of "trading on our race and class and outsider status in order for everybody to get a hearing. That is, people were more likely to listen to us tell the stories of the residents than the residents themselves. And yet, everybody agreed that it was important to provoke those conversations and then set up mechanisms on the inside so that those conversations could happen organically rather than be artificially initiated by outsiders."

But the search conferences and workshops and story-telling rituals of action research processes are designed in the light (or darkness) of encompassing power relations. As deliberately diversified small groups meet to discuss problems and opportunities, participants can learn in surprising ways. The same happens in mediated negotiations or joint problem solving processes: participants find themselves meeting together with supposed adversaries from their own city, adversaries with whom they may have never spoken informally, and they can learn about one another in ways that do not blunt their objectives, do not co-opt them, but make them capable of serving, elaborating, and focusing their interests more effectively. Writing of the challenges of designing deliberative processes in the public sector, Morten Levin writes,

> "Everyday life in small municipalities does not grant legitimacy for such close cooperation. . . . Designing a process of discovery and triggering an

> active push for involvement is required to bring about (effective public-private cooperation). . . . Creating legitimate fora for mutual exploration, development, planning, and action seems to have allowed the parties to begin to engage in worthwhile joint activity."

Rather than finding themselves scavenging in an organizational "garbage can" for idiosyncratically interesting objects to work with, PAR participants may find themselves learning in surprising and unpredictable ways as they participate in loosely goal-directed, but ritualized performances of sharing stories together, brainstorming possibilities, listing strengths and weaknesses of salient organizations and so on. These ritual processes enable them to listen and learn without forcing their attention so narrowly that they miss the richness of concerns and capacity that others bring to their encounters. In such ritualized meetings, participants are asked to listen more than to choose, to consider new information and feelings more than to accept or reject them.

So too, re-constituting political space -- trying to "create a forum where people could begin to address class and race issues," Fine tells us that she saw her job as "collecting these stories and feeding them back to a mixed group so that people could hear each other's stories."[25] Mindful of power relations, she and her colleague suggested ground rules for the shelter's discussion too: "people shouldn't be saying whether or not somebody else's perspective is true or not, but instead how they hear it, how they feel about it, what it means for the organization." Fine and her colleague encouraged not so much an informative exchange of arguments here but a transformative ritual of deliberation. In this structured conversation, issues of action ("how they hear it"), emotion ("how they feel about it"), and consequences ("what it means for the organization") might all be explored.

7.11 Learning from structured complexity: rituals as aids to dialogic and deliberative rationality

So the ritual richness of participatory processes can re-mind participants of significant issues by bringing before them - enabling them to perceive, to recognize, to re-member, and even to express (re-cite) - items of value they may have not had immediately or clearly in mind. Loosely goal-directed rituals, then, enable them to learn about and to acknowledge value, to identify new concerns and issues, without forcing choice.

These ritual encounters allow participants to listen and learn, to watch and see, to feel and identify concerns they were not clear about earlier. But these encounters also involve and partially commit participants as they provide occasions for their own performances of story-telling and issue identification. Acting in front of others on such ritual occasions, participants take certain stances of tone and posture and attentiveness. As they tell their stories, their stories tell a good deal about their selves too.

This much leads to a curious result. If the richness of ritualized story-telling and listening exposes participants to details, to particulars that re-mind them of their own concerns, then the more narrowly they focus their meeting upon their immediate purposes, the less they may learn in ways they themselves need.[26] Their own intensity of concern and their presumptions about others -- and the information they do not yet have -- hold them captive, prevent them from exploring issues and options that would help them. If curiosity kills cats, presumption kills negotiators. Their narrow purposiveness can not only blind them, but it can blind others to the range of values they might probe together, if not fully serve.[27] This is, more generally, the argument for creativity and play in complex, uncertain, and ambiguous decision-making contexts.

But the "richness of ritualized story-telling and listening" involves not just any ritual form. The ritual structure provides the vessel but not the message. The stories of participants' concerns can be expressed in a variety of participatory settings, but the more narrowly pre-scribed, expressible in advance (pre-dictable) they are, the less informative, demonstrative, evocative, refreshing, and even redemptive they will be. If participants are to be able to acknowledge one another's concerns, they must be able to address one another. If they are to be able to address their differences, they must be able to listen to one another and to speak responsively as well.[28]

Consequently, the danger arises: the thinner the ritualized occasions that might enable the expression and exploration of specific concerns and interests -- be these formally structured and time-limited presentations or relatively informal conversations over drinks or meals (or coffee breaks, break-out groups,corridor conversations, or bar-room or bathroom chats) -- the less will be learned. The challenge here for planners and PAR designers is the structuring of deliberate distraction, the deliberately enabling distraction from participants' own narrowness of focus that prevents them from remembering, recognizing, and learning in ways they wish to.[29]

The more purposeful and simple our statement of "positions," to use the formulation of advocates of interest-based bargaining, the more meagre our opportunities to learn, to probe new options, and to design strategies that will effectively serve our ends.[30] Once we understand that we begin deliberative discussions with limited rationality, less information about others, and still less information about the full range of possible agreements we might devise, we can then appreciate that we need not so much bottom-line simplicity in our deliberations but rather a ritually organized, optimum complexity of description and narrative richness that enables us to learn -- and, of course, falls far short of chaos and immobilizing confusion.

Such ritually organized complexity honors the participants' senses of direction but not any single purpose; it enables the presentation and articulation of clusters of concerns and interests rather than narrower positions; it allows the consideration of relevant questions before narrowing participants' attention to any over-riding one. In this light, the ambiguity of interests and priorities, even of law and political mandates, is an opportunity for participants to explore and work to resolve together rather than only a source of indecipherable confusion.

The "search conference" is one such ritual design that generates both intentional complexity and deliberate distraction: it sketches a broad purpose, a direction of commitment to be explored together; it invites participants to teach one another about relevant history, facts, techniques, and opportunities. In doing so, participants can learn not only about strategies to serve their interests, but also about one another and about interests of their own which were less salient to them at the beginning of the search process.[31]

But more happens here too: as the ritualized groupings and conversations occur with conventions and rules of their own (even those of ordinary small-talk in a coffee-break), participants find themselves not only bound by these rules but enabled by them as well. Not only are certain forms of formal argument and posturing precluded, at times, but other forms of listening and being heard become possible in these new arenas of meetings of parties and in some cases, previous disputants. So too, do seductive information technologies (interactive computing, multi-media systems, and sophisticated mapping programs, for example) pose both threats and opportunities here, for different forms of representation enable different forms of interaction and deliberation, different forms of emotional responsiveness, memory, insight, and perception (Peattie 1991).

Looking narrowly and abstractly at the "tough choices" before them, PAR participants might make choices (and treat each other) more callously than they must, if they let themselves ignore important values because they cannot serve them at the moment. But here participatory rituals that ask all participants to step back before going forward, to look broadly before focusing narrowly, can help. Ritually structured re-minders of the values at stake in their actions can help to keep participants honest, help to keep them to the standards they respect, help more practically to ensure that their limited attention does not blind them and others to the necessary "costs" or "losses" or "downsides" of their own decisions.

7.12 Acknowledging others: encouraging a politically deliberative community

All this may help us to understand that decision-making involves not only cognitive choice but social expression, social articulation that divisively or redemptively addresses those served and those disserved, winners and losers alike. So may election winners, for example, often acknowledge in acceptance speeches the virtues of their defeated opponents and recognize the presence and significance of the opposition. So too can the explanation of a political decision acknowledge and credit opposing arguments, not ridicule or ignore them, even as it justifies and seeks to motivate and gain support for another course of action.

Such public and articulated acknowledgement of conflicting and pressing values does not solve a problem; it works ritualistically to re-build relationships and to prepare the social basis for future practical action. It works to reassure losing interests of their standing; it works to transform its audience from a collection of winners and losers on one particular issue to a more integral and enduring political community of members respecting one another's differences.[32]

Such ritualized acknowledgement of multiple and conflicting values, then, might counteract even a coming decision's potential divisiveness not only by re-minding its audience of the significant concerns of opponents but by encouraging all parties to re-member their common commitments to a political life together. So Fine's challenge is help the women's shelter not to avoid choice but to maintain the solidarity that will allow staff and clients to make difficult choices and continue together afterwards.[33] Reardon speaks too of the importance of participants' being recognized, of their initial "skepticism" giving way to confidence:

> "The fact that they are now getting recognized really drives it home to them because you're talking about people who have never had a chance to articulate their concerns to people in power. As the organization begins to build, develop, and grow, it has the ability to offer these people that opportunity that's very important to all of us: to be heard. . . ."

Counterbalancing their participants' own limited attention to their pressing purposes, then, rich rituals that allow a rehearsal of concerns, commitments, and values can help their participants to learn in ways they are unable to intend, to foresee, or to predict. In choosing this or that ritual of story-telling together, participants choose a more hierarchical or conversational way of learning together, of shaping solidarity in different ways, along class lines or not, lines of expertise or not, and so on. But what they will learn in the conversation cannot be foreseen with much confidence in advance.

Into the ritual occasion of sharing stories and concerns, for example, or sharing lists of strengths and weaknesses, threats and opportunities, come concerns and relationships. With the concerns come particulars and facts that matter, details suggesting issues to be explored. With the relationships come evolving possibilities of understanding, of mutual agreement and contingent promising, of collaborative opportunities, of going on together in unforeseen ways. Participatory and deliberative processes, then, are not simply sites of bargaining and trading, they are occasions on which participants, their senses of self and opportunity, and their practical relationships can be deliberately, if gingerly, transformed.

7.13 Conclusion: the significance of ritual in participatory and deliberative settings

So what? Consider seven points quickly.

First, the analysis of learning through deliberative, participatory rituals suggests that we learn not only with our ears but with our eyes and hearts. We learn not only from surprising information that leads us to propose new hypothetical lines of action to test, but we learn from style and passion and allusion too. We learn to reframe our predictions and strategies, but we learn to develop new relationships and even senses of ourselves as well. In participatory processes, we not only generate arguments, but we construct networks and new organizational forms as well.

Second, this analysis suggests that participatory groups learn in a great many ways. We must be wary of focusing so much on argumentative learning that we fail to appreciate participants' learning of skills and confidence, appreciation of and respect for others.

Third, this account might help us improve upon the intriguing but limited "garbage can" model of decision-making in which solutions find problems, choices find preferences, and decision-makers find work. Transformative rituals enable the deliberative reconsideration of preferences and choices, of problems and solutions, of decisions and decision-makers, by enhancing participants' limited attention, information, and rationality.

Fourth, an analysis of transformative and ritualistic learning in participatory settings can help us to understand the significance of the very messiness, complexity, detail, and moral entanglements of living stories and dramatic role-playing presentations. Far from being simply diffuse or emotional, these stories are accounts of value and identity, of abiding concern and complexities that are ignored at practical risk.

Fifth, this account can help us to understand that the selective representation, in speech or writing, story-telling or role-playing, of problems and choices and decision options always has political and moral resonance, as it acknowledges or fails to acknowledge the depth of human concerns at stake. Social theorists write of the redemptive, value-acknowledging character of rich narratives because they are concerned about the ways that history or literature may do justice, or fail to do justice, to the lives represented there. Members of participatory groups surely have no less obligation to respect the pain or loss that their actions will cause as they make what too easily get called "tough choices."

Sixth, this account might help us to understand the oblique and surprising, non-intentional learning we must protect and even enable ourselves to do in practical situations. We know that we do not know in advance what we will come to value and what we need to learn about others and the complex political environments we face. Knowing that, we need not only to offer and test hypotheses, to move and consider the "back-talk" of the situation as Don Schön argues, but we need to look and listen, to meet others and recall commitments, to allow ourselves to remember values less of the moment but no less to be honored, and to resist what Robert Coles so wonderfully called "the rush to interpretation."

Seventh, finally, the analysis of ritualized learning helps us to resist the temptations of economistic bottom-lines, and so to understand that political deliberation, the participatory attempt to reconsider ends and means together, to learn about what we want and what we can do, is far richer than a process of bargaining and trading across pre-given "interests" that we have in any stable and well-defined rank orderings. The transformative ritual account of participatory or deliberative learning helps us to understand that in participatory processes we may not only argue to test strategies for maximum gain, but we may make and act on new agreements, transforming ourselves and others in the process. We generate new ideas, but also new organizational forms. We develop better practical hypotheses, but also new capacities of action and cooperation. We may transform ideas and ourselves, in unexpected but valued ways, as well.

References

Argyris, C. and D. Schön (1978) Organizational Learning, Addison Wesley, Reading, Mass.

Boyd White, J. (1985) "Law as rhetoric, rhetoric as law: The Arts of cultural and communal life", U. Chicago Law Review 52 (3), 684-702

Brown, L.D. and R. Tandon (1983) "Ideology and political economy in inquiry: Action research and participatory research", J. Applied Behavioral Science 19 (3), 277-294

Cavell, S. (1976) Must We Say What We Mean?, Cambridge U. Press, Cambridge

Cavell, S. (1993) "Must we mean what we say?, and Ricoeur", in Levin, cit.

Cohen, M. D., J. G. March, and J. P. Olsen (1988) "A garbage can model of organizational choice", in J. March, Organizations and Decisions, Basil Blackwell, New York, p 296.

Dewey, J. (1922) Human Nature and Conduct, Henry Holt1930, New York

Edelman, M. (1964) The Symbolic Uses of Politics, U. Illinois Pr, Urbana

Fals-Borda, O. and M. Rahman, eds. (1991) Action and Knowledge: Breaking the Monopoly with Participatory Action Research, Apex Press, New York

Fine, M. (1993) "Inquiring institutions: Transformation through critical inquiry", in Forester, Pitt and Welsh, cit.

Fisher, R. and W. Ury (1983) Setting to Yes, Penguin, New York

Fiskin, J. (1991) Democracy and Deliberation, Yale U. Press, New Haven

Forester, J. (1992) "Envisioning the politics of public sector dispute resolution", in S. Silbey and A. Sarat, eds. Studies in Law, Politics, and Society 12, 83-122

Forester, J. (1993) "Learning from practice stories: The priority of practical judgment", in F. Fischer and J. Forester, The Argumentative Turn in Policy Analysis and Planning, Duke University Press, Durham

Forester, J., J. Pitt and J. Welsh (ed.) (1993) Profiles of Participatory Action Researchers, Cornell University Department of City and Regional Planning, 191-page typescript, Ithaca

Freire, P. (1970) The Pedagogy of the Oppressed, Seabury Press, New York

Garcia, A. (1991) "Dispute resolution without disputing: How the interactional organization of mediation hearings minimizes argument", Amer. Soc. Review V (56), 88-835.

Gaventa, J. (1991) "Toward a knowledge democracy: Viewpoints on participatory research in North America", in O. Fals-Borda and M. Rahman eds., cit.

Gaventa, J. and B. Horton (n.d.) "A citizen's research project in Appalachia, USA", Convergence 14 (3)

Greenwood, D. (1991) "Collective reflective practice through participatory action research: A case study from the Fagor cooperatives of Mondragon", in D. Schön (ed.), cit.

Gutmann, A. (1993) "The challenge of multiculturalism in political ethics", Philosophy and Public Affairs, 22 (3), 171-206

Habermas, J. (1975) Legitimation Crisis, Beacon Press, Boston

Healey, P. (1993) "The communicative work of development plans", Environment and Planning B: Planning and Design 20, 83-104,

Horton, M. and P. Freire (1990) We Make the Road By Walking, Temple U. Press, Philadelphia

Isaacs, W. (1993) "Taking flight: Dialogue, collective thinking, and organizational learning", Report of the Center for Organizational Learning's Dialogue Project, MIT

Kertzer, D. (1988) Ritual, Politics, and Power, Yale University Press, New Haven

Levin, M. (1993) "Creating networks for rural economic development in Norway", Human Relations 46 (2)

Lukes, S. (1975) "Political ritual and social integration", Sociology 9 (2), 289-308

Meyerhoff, B. (1982) "Life history among the elderly: Performance, visibility, and re-membering", in J. Ruby (ed.), cit.

Moore, S. and B. Myerhoff (1977) Secular Ritual, Atlantic Highlands, Humanities Press, New Jersey

Murdoch, I. (1970) The Sovereignty of Good, ARK, London

Neustadt, R. and E. May (1986) Thinking in Time, Free Press, New York

Nussbaum, M. (1986) The Fragility of Goodness, Cambridge U. Press, Cambridge

Nussbaum, M. (1991) Love's Knowledge, Oxford U. Press, Oxford

Peattie, L. (1987) Planning: Rethinking Ciudad Guayana, U. Michigan Press, Ann Arbor

Pitt, J. (1993) "Beyond rules and theory: Linda Stout's community organizing practice", CRP, Cornell U., typescript.

Reardon, K. et al. (1993) "Participatory action research from the inside: A profile of Ken Reardon's community development practice in east St. Louis", The American Sociologist 24 (1), 69-91

Rosen, M. (1985) "Breakfast at Spiro's: Dramaturgy and dominance", J. of Management 11 (2), 31-48

Ruby, J. (ed.) (1982) A Crack in the Mirror: Reflexive Perspectives in Anthropology, University of Pennsylvania Press, Philadelphia

Schön, D. (1983) The Reflective Practitioner, Basic Books, New York

Schön, D. (1991) The Reflective Turn, Teacher's College Press, New York

Schön, D. (1992) "The theory of inquiry: Dewey's legacy to education", Curriculum Inquiry 22 (2), 119-139.

Turner, V. (1969) The Ritual Process: Structure and Anti-Structure, RKP, London

Turner, V. (1974) Dramas, Fields and Metaphors: Symbolic Action in Human Society, Cornell U. Press, Ithaca

Viggiani, F. (1991) Ph.D. dissertation, Dept of Industrial and Labor Relations, Cornell University

Notes

[1] See, e.g. John Dewey (1922), Chris Argyris and Donald Schön (1978), Paulo Freire (1979), Jürgen Habermas (1975). For a discussion of the ideological dimensions of applied research, see L. David Brown and Rajesh Tandon (1983). Cf. Davydd Greenwood (1991), John Gaventa (1991), and Don Schön (1992).

[2] This and all subsequent quotations of Ken Reardon come from: John Forester, Jessica Pitt and John Welsh ed. (1993). Cf. Ken Reardon, John Welsh, Brian Kreiswirth, and John Forester (1993); John Gaventa and Billy Horton (n.d.).

[3] On the consideration of fundamental disagreements in deliberative processes, see Amy Gutmann (1993), p.199. Gutmann writes, "Deliberation may sometimes increase moral conflict in politics by opening up forums for argument that were previously closed. . .Deliberation encourages people with conflicting perspectives to understand each other's point of view, to minimize their moral disagreements, and to search for common ground, but it begins by opening politics up to a range of reasonable disagreement that is restricted by less deliberative politics." 199.

For rich accounts of many forms of participatory learning, see Myles Horton and Paulo Freire (1990).

[4] For recent theoretical work on dialogue still concerned with epistemology more than with ethics, with meaning and assumptions more than with speech and performance, with the nature of underlying paradigms more than with rituals of contingent promising, see William Isaacs (1993). On laboratories for judgment, Peter Koschitz writes aptly that "by assessing the possible outcomes of a measure meant to solve a certain problem we can learn much more about what we value than by theoretically discussing value priorities." (personal correspondence, 16/9/93, Swiss Federal Institute of Technology, Zurich).

[5] U.W. personal conversation, Mimrebu seminar, Norway, October 1993.

[6] Cf. All quotations from Gonzalez are taken from personal communication (e-mail), October 1993. The research was reported internationally in a paper to the XII World Congress of Sociology in 1990, and in an article in the Journal of Social Issues, V46, 1990, authored by Briceno, Gonzalez, Phelan and Ruiz, Laboratorio de Investigaciones Sociales, Universidad Central de Venezuela, 1984-1987.

[7] On "the moral relevance of surprise" and the priority of recognizing unique particulars in deliberation, see Martha Nussbaum (1991), p72.

[8] See Morten Levin (1993), pgs 193-218. All subsequent quotations from and references to Levin come from this source.

[9] Cf. Barbara Myerhoff (1982), p.111: "To signify this special type of recollection, the term, "Re-membering" may be used, calling attention to the reaggregation of members, the figures who belong to one's life story, one's own prior selves, as well as significant others who are part of the story. Re-membering, then, is a purposive, significant unification, quite different from the passive, continuous fragmentary flickering of images and feelings that accompany other activities in the normal flow of consciousness. . . .A life is given a shape that extends back in the past and forward into the future. It becomes a tidy edited tale. Completeness is sacrificed for moral and aesthetic purposes. Here history may approach art and ritual. The same impulse for order informs them all.".

[10] See Michelle Fine (1993). All references to Fine, and quotations below, come from this source.

[11] Reminding may have a special philosophical and practical significance, for in attending newly to words we can attend newly to the world, as Stanley Cavell (1976) suggests at p20-21,.

[12] Attending to objects, we can reconsider roles and our own possibilities of action in the world of those objects: the soccer ball (I can play just a bit), the car (I have little time or money to travel), the radio (I enjoy, and look forward to, listening to the blues) and so on. So: rituals occasion stories, stories re-present significant objects; significant objects give us regulative and constitutive norms and roles; norms and roles give us ways of going on separately or together, i.e. fluid social identities; identities give us a sense of possibility (and limits) and relevant questions to explore regarding related concerns (Could I get time off to travel with the car?)...

[13] This helps us to see why people often say they "love" or "hate" ritual. In both cases, ritual participation is understood to shape identity: those who "hate ritual" understand that their own ritual participation presumes to make them something, someone, they do not want to be. Ritual haters often express their dislike of hypocrisy: that the identity affirmed and presented in the ritual performance is not genuine. Those who "love ritual" often find ritual participation a means of honoring what they cannot express, a means of being part of a larger community of practitioners with a history of beliefs, aspirations, values, and commitments that again finds poorer expression in ordinary language than in the pomp and circumstance of highly sedimented, historically resonant ritual practices. All of which is not to make a case for or against any particular ritual...but to appreciate that ritual performance shapes social identity, which both lovers and haters of ritual understand. More generally, if we value X, we want to participate in the rituals associated with X. Conversely, we resist participating in rituals associated with what we do not honor because those ritual performances make us someone we are not, makes us duplicitous, and publicly so.

[14] In public policy settings we often find the rituals called "formal public hearings." These are clearly structured, even deliberately stilted forms of public presentation that channel the stories of participants to a formal hearing board and prevent participants from conversing and forming any joint agreements among themselves. Such public hearing rituals often seem to exacerbate conflict because they encourage exaggeration and posturing while they discourage clarification of issues, testing of claims, and exploration of options or potential agreements. These ritualized "hearings" constitute their participants not as citizens who must speak and come to terms together but as poker playing adversaries, as masters of the bluff rather than as good listeners able to respond to one another.

[15] See S. Lukes (1975), p.302. Cf. Victor Turner (1969), David Kertzer (1988). For an integrative view, see Michael Rosen (1985). See also Sally Moore and Barbara Myerhoff (1977), Victor Turner (1974), Murray Edelman (1964).

[16] The emotional texture of PAR and deliberative processes appears to be a seriously neglected area of inquiry. Cf. Martha Nussbaum (1991), Chapter 2, p81. Cf. John Dewey (1922).

[17] Cf. James Fishkin (1991, p.29) on the significance of democratic deliberation: "Without deliberation, democratic choices are not exercised in a meaningful way. If the preferences that determine the results of democratic procedures are unreflective or ignorant, then they lose their claim to political authority over us. Deliberation is necessary if the claims to democracy are not to be de-legitimated.".

[18] Again then, we learn about such value allocations by evaluating others ritual performances: so someone means to help but does so ineptly, or callously (bedside manner), and we see that in addition to intention, value is allocated. By attending to ritual performance, we can not only respect (more or less) and recognize the person, but we can assess the way they shape or obscure or respond to or illuminate or threaten parts of the social world we share (particular relationships, or next week's meeting).

[19] See S. Cavell, (1993) on the "surplus of meaning".

[20] Cf. on democratic rituals, Frances Viggiani (1991)

[21] On the importance of creating "safe spaces for story telling" in participatory deliberations, see also Jessica Pitt (1993).

[22] See Michael D. Cohen, James G. March, and Johan P. Olsen (1988)

[23] C.F. J. Forester (1992), and Patsy Healey (1993).

[24] Personnel communication with the author, October 1993.
[25] See Forester (1992) and Angela Garcia (1991)
[26] We shape the allocation of value in the world in ways we do not intend, and we can view others' actions as similarly shaping value in non-intentional ways. So we say, for example: 1) "I know you didn't mean to hurt her feelings, but you did," (regarding the value of another's integrity), or 2) " I know you didn't mean to disclose the secret" (regarding the value of particular information in a relationship), or 3) "I know you didn't mean to get us into this trouble" (regarding the value of a violated norm), or 4) "I know you didn't mean to break the computer/glass" (regarding the value of a valued object) and so on. This suggests that we commonly evaluate our own and others actions as they affect value in the world beyond whatever we or others intend.
[27] There are issues here of tragic limits of attention in complex decision situations. We may care about far more than we attend to at the beginning of a negotiation, for example, and the complexity of decision-making and participatory rituals may enable us, accordingly, to acknowledge more (or less, depending on their form and content) of those concerns. Cf. Chapter Two of Martha Nussbaum (1986) and its related arguments against the neglect of loss encouraged by facile beliefs in the commensurability of value concerns. The leader who ignores the losses caused by his or her "correct" decisions may be not only blind but blinding, neglecting important values and leading others to fail to appreciate them as well.
[28] Seen in this light, public hearings are virtually pathological rituals that minimize responsive interaction and maximize exaggeration and adversarial posturing.
[29] There is a prisoner's dilemma analogy to be made here -- going for the sure thing, the interest at hand, we fail to see that we have still other interests whose mutual satisfaction will serve our interests better still. Pre-commitment (putting the alarm clock out of our reach so we can't go for the short-run desirable sure thing) is one response: pre-commitment not just to outcomes (Ulysses binding himself to the mast to resist the Sirens) but to processes of listening, consideration, reminder, articulation of memory, exploration of options, delay of premature agreements, involvement of third party intermediaries, testing of evidence, probing of mutual contingent promises and offers, and so on.
[30] See here the extraordinarily influential and deceptively simple little book by Fisher and Ury (1983). Compare Richard Neustadt and Ernest May's dictum about decision diagnosis: "Don't ask, "What's the problem?" Ask, "What's the story?" That way you'll find out what the problem really is." (Neustadt and May 1986, p. 274).
[31] See Levin, cited above, eg.
[32] See Boyd White (1985); for broader arguments about the practical politics of story-telling and the development of political judgment, see the essays in J. Forester, Politics, Pragmatism, and Vision (being prepared for publication) and Seymour Mandelbaum, The Poetics of Open Moral Communities (being prepared for publication).
[33] Cf. Davydd Greenwood's work on Mondragon, arguing that "deliberation does not resolve conflict necessarily. The consensus on the rules creates situations in which it is possible to explore in detail the fundamental differences of view of the participants. In the process, the trivial differences can be sandblasted away, leaving the major differences of vision or value commitment that underlie them. Once this has happened, the situation becomes political in a new sense. Gone is the confusion. What is left is an attempt to mobilize constituencies in support of one vision or another and the winners and losers remain loyal opponents." (personal comm. 1/10/94).

John Forester
Faculty of Architecture and Town Planning
Cornell University
USA

Technion Department of City and Regional Planning
Haifa 32000
Israel

8 ASSESSING THE POLITICAL DIMENSION OF STRUCTURE PLANNING PROCESS

Abdul Khakee

8.1 Introduction

The notion that urban planning is a political process has its origin in the debate in the late 1960s. A major reason for this was the frustration with traditional planning that had very little impact on urban development. It was realised that if planning was to have a real impact it should be concerned with implementation issues (Catanese, 1974). Implementation involves very real political issues like resource mobilization, feasibility and conflict resolution.

The shift in emphasis in planning research (and for that matter even in planning practice) from plan to planning process has to do with the increasing recognition of the political nature of the process. In focusing on the plan-making exercise, researchers found it easier to understand not only the forces that ultimately determine the outcomes of the exercise, namely plans and policies, but also whether these outcomes were in accord with goals and objectives of politicians who would eventually effectuate them (Khakee and Eckerberg, 1993).

Despite the fact that the shift in focus is a quarter of a century old, very little research has been carried out to evaluate the planning process itself (Masser, 1983). Much of the research has focused on the impact of planned intervention. In this context the relevant question if it is at all feasible to use the term "evaluation" in determining the results of a planning process that is to a great extent a negotiating process, a dialogue involving professional planners, politicians, organised community interests and the public (Healey, 1992). Besides the production of a plan, this process results in several behavioural, organizational and policy changes that may be unrelated to any prior goal or to several. Thus it may be difficult to speak of "evaluation" of a process in the sense that the intention is to find out specified sets of results in relation to one or several goals. It may be useful to speak of "assessment" of planning process since the latter may be affected by more than the stated intentions of planning legislation.

Previous research into evaluation methodology is of slight help in assessing planning processes. A comprehensive assessment methodology has to cover a wide variety of

D. Borri et al. (eds.), Evaluating Theory-Practice and Urban-Rural Interplay in Planning, 105–116.

aspects of the planning process. One such attempt (Khakee, 1994) classifies these aspects into three dimensions of planning: the normative dimension that covers the scope and legitimacy of planning, the methodological dimension that covers procedures for applying knowledge to action during stated periods of time, and the institutional dimension that covers the means of managing and co-ordinating planning activity.

The aim of this paper is to assess the political dimension of the structure-planning process in Swedish municipalities after the enactment of the Planning and Building Act in 1987. The paper focuses on the methodological requirements for such an assessment as well as the results of applying a given assessment approach. Following its introduction, the paper comprises four sections: first, the political nature of the structure plan as defined by the planning legislation, second some theoretical issues of the politics of planning as they are relevant in the assessment exercise, third some findings of empirical research, and fourth and finally, some reflections on the need to develop a comprehensive methodology for assessing the planning process.

8.2 The political nature of the structure plan

The structure plan, which replaces master-plan and municipal-planning guidelines of previous planning practice, is a local development plan. The word "local" needs to be emphasized since the municipalities are no longer required to submit the structure plan to the county administration for approval and to the central government for ratification (in fact, very few master plans were ever ratified). The structure plan has to be approved by the municipal council, which emphasizes its political status. A relevant aspect in this context is that the municipality need not implement its plan: its primary role is to provide guidelines for future development and land conservation.

The planning legislation requires that the structure plan contains an account of public interest and how it is considered in connection with decisions on land use and water resources. It further specifically requires a municipality to account for how planning pays due attention to national interests as outlined the Natural Resources Act, which was enacted at the same time as the Planning and Building Act and is a further development of the national guidelines for the use of land and water resources approved by the parliament (Ödmann, 1992).

The legislation also emphasises the legitimacy of planning by requiring each municipality to consult citizens twice during the planning process: firstly, when the preliminary plan proposal is sent out for comments to various organised interests, and secondly when the final version of the plan is put on exhibition, when they may again record their comments, which must be added as an appendix to the structure plan.

The plan approved by a municipality becomes the basis on which mandatory area action regulations and detailed development plans are prepared. As to building controls, the new legislation replaces the general prohibition on building in the planned area with a provision that each proposal be judged on its merits. Building permission may no longer be withheld on formal grounds. This means that development involves negotiations when public interest comes in conflict with private development interests (Floderus, 1986).

These developments towards greater local autonomy and changes in the role of municipal planning have taken place while the role of the public sector and especially the municipal monopoly on the production of various services have been called in question. The important point at hand is that while a municipality may include other than mandatory matters relating to national interests, it must submit the plan to public comment and must include such comments in the final version of the plan that is to be approved by the municipal council. This statutory requirement of democratic support makes the plan a political instrument for community development.

8.3 Theoretical premises

Two major themes emerge from recent literature on the politics of planning: planning in the wider system of power and control, and planning as a professional activity (Benveniste, 1989; Low, 1991).

In a wider system perspective the legitimacy of planning is of course crucial and has been debated as long as the issue of political intervention in society has been topical in political and economic discourse. Most recently, there have been several important debates on this issue. During and after the Great Depression of the 1930s, the debate was about the compatibility of market economy and central economic planning (see, for example, Eckstein, 1971 and Friedmann, 1987). The post-war French experiments in central economic planning aroused a debate on the usefulness of indicative planning in a market economy (Lutz, 1969; Khakee, 1978). In the last few years, debate has focused on the role of public planning in face of the challenge of the market. Neo-liberal criticism as propounded especially in thatcherism has been an important component in this (Thornley, 1991).

While this debate has addressed the legitimacy of all forms of public planning, it has focused on national planning. Some researchers e.g. Healey (1989), argue, however, that there has been a greater understanding that land-use and environmental planning has been needed more than other forms of public planning and point particularly at the role of public planning in redressing land-market failures and promoting the requirements of nature conservation. While this argument seems to have been accepted by even by the most ardent critics of public planning (Thornley, 1991), Low (1991) extends it to distributive aspects in writing that urban planning was introduced "both because the market failed to take account of longer-term issues and the interests of large constituencies (all who have an interest in the efficient functioning of the city), and because many people had insufficient market power to command satisfaction of their needs" (p. 30). Low summarises here an opinion held by urban-planning researchers and practitioners. Urban planning has a strong component of social reform. Redressing the ills of urban growth - overcrowded housing, poorly maintained buildings, inadequate provision of public services - is central to it. Thus some normative concept of the public interest must underlie all planning activity.

In emphasizing the possibility of conflicts between national and local interests, Swedish planning legislation accepts pluralistic interests instead of a sole public interest. It also points to the requirement of arriving at a compromise between different interests but, while once discussions between national sectorial agencies, county administrations and municipalities achieved this end, it no longer lays down how compromise should be reached. The relevant questions are how the entire issue of public interest was perceived during the structure planning process and, given the framework of existing legislation, what planners perceived as threats to the public interest? What were the major points of conflict between national and local interests? In what way were the conflicts structured? What framework of negotiations was adopted in order to negotiate on public interests?

Local-government processes are enmeshed in a complex political web, and the power of a local government to plan intersects with other sources of power. The limitations on its power became obvious at the end of 1960s, when many of the worse features of urban growth became apparent and led to demands for public participation in planning. Advocacy planning was one response to this demand. But, as Low (1991) puts it, this was inadequate since public participation implies new institutional rules, which advocacy planning does not do. Democratic support for plans requires public control that goes beyond local representative government.

Public participation is one of the most debated issues because in no other area of public-decision making are principles so different from practice. Research shows that planners and decision makers favour public participation in principle but, for several reasons, oppose it substantially in practice (Bohm, 1985): it is seen as interfering with a well-ordered process, as a putative cause of instability, and as a way for the articulate unintentionally to prevent the inarticulate from expressing their preferences.

A major reason for the lack of a satisfactory means for public participation is not a lack of satisfactory methods but because the political system sets limits to it. In countries like Sweden, the belief in representative parliamentary forms of government is so strong and pervasive that there is a deeply rooted aversion to extra-parliamentary forms of action (Miller and Kraushaar, 1979). Public participation is often regarded as extra-parliamentary. At the same time several studies have revealed a gap between planners' and politicians' values and those of members of the public. As time elapses after elections, politicians tend to align themselves more and more with local-government officials, and there is much evidence that professional, not general, socialization has an impact on the values held by politicians. (Khakee and Dahlgren, 1990).

Therefore some normative concept of democratic support must underlie all planning activity, and any assessment of the planning process should include issues like planners' perceptions of public participation, of restrictions on it, trade off between democracy and efficiency, and methods appropriate to assessment.

Planning is no longer regarded as a technocratic exercise in rational analysis by an isolated cadre of professional planners. This change in how planners' roles are seen has occurred not only because planning as an activity independent of political process did not work but also because there is little evidence that planning ever was an isolated area of bureaucracy (Vasu, 1979; Howe and Kaufman, 1981; Khakee and Dahlgren, 1990). Planners' active involvement in implementation no doubt increased once they realized the only way to ensure the success of their plans was to work closely with politicians.

The extent of planners' influence in politics depends on how they organize their work, what responsibility and authority they exercise, how they resolve conflicts, establish networks, build coalitions, gather resources and manage uncertainty (Benveniste, 1989). But planners' mediation in urban development is not neutral and value-free, being based on the degree of their commitment to equity and other values that is evidenced from the set of activities which they are typically involved: "promoting and regulating industrial and commercial development; facilitating the development of public services ...;

redevelopment of existing residential areas; 'slum clearance'; 'comprehensive development'; 'gentrification'; converting rural land to urban use and, in so doing, placing density controls and other restrictions on the use of land; ... conserving historic buildings and urban places, and rural and natural landscapes; enforcing compliance with planning regulations" (Low, 1991, p. 29). In order to carry out these activities, planners must apply not only scientific methods but also their personal values. Thus the relationship between objective scientific knowledge and subjective values become crucial in planning practice.

In assessing the structure-planning process, relevant questions include whether it took account of rationality considerations; was incremental; included to any extent questions of implementation; accorded importance to given modes of knowledge (projections, quantitative estimations, local opinion polls, and so on); managed participation, group processes, and uncertainty satisfactorily; and whether this process caused 'lessons to be learned'.

8.4 Empirical analysis

Three sets of studies were designed to assess the structure-planning process. The first comprised a qualitative and quantitative evaluation of structure plans with special emphasis on how they reflect the planning process; the second a questionnaire survey of some selected municipalities; and the third in-depth interviews with planners in five municipalities to elicit their own evaluations of the planning process.

he qualitative and quantitative evaluation was made of the structure plans submitted by 254 (of 286) Swedish municipalities. The 32 that submitted no plan did so because either a plan had not been approved by the council or commented on by all community interests. After our preliminary perusal of the 254 plans, we developed 24 indicators of how accessibly plans presented their contents; and covered various substantive areas, different types of conflict, and consideration of long-term issues. As we scored each indicator, a plan could score a maximum of 55 points. In the event, scores ranged from 9 to 37 points and, on the basis of this spread, we divided plans into five categories: (1) 50 'simple plans' (9-13 points) that presented maps and a limited amount of analytic material; (2) 95 'plans with narrow perspectives' (14-19 points) presenting almost entirely land-use issues; (3) 77 'plans with relatively broad perspectives' (20-26 points) covered many substantive issues but with a superficial analysis; (4) 23 'ambitious plans' (27-32 points) that broadly covered all or nearly all the substantive issues, and (5) 9 'thorough plans' that treated all issues comprehensively.

The questionnaire survey covered four municipalities in each of these five categories (20 in all). Of the 133 local-government officials and politicians whom we had identified as having major responsibilities for preparing plans and to whom we sent the questionnaire, 86 (65%) answered. A major reason for the drop out was that many had left local-government service or, having become external consultants, did not feel "obliged" to answer.

The questionnaire included 45 questions under the three general heads of the political, methodological and organizational dimensions of planning. The questions focused on the variables selected for each dimension and, knowing many planners are reluctant to give extensive written answers, we posed most questions in multiple-choice form.

In-depth interviews with planners in a municipality in each of the five categories (5 in all) not only involved their own assessments of how the process was carried out but also of the major aspects of the politics of planning namely public interest, public participation, the role of planners and the relationship between knowledge and values.

8.5 Politics of planning

Legitimacy of planning

The pursuit of the public interest provides legitimacy for planning. The existence of a sole public interest on which planning can be based has been a major issue in planning research (Vasu, 1979). With the advent of advocacy and social-planning theory, research interest shifted to the implications of pluralism (Davidoff, 1965). Can planning by a local-government authority be justified if it leads to conflicts between different public interests? Supposing such conflicts arise, how can such an authority resolve them? An extension of the pluralism debate is the realization of conflicting national and local public interests. In the case of Swedish structure planning this conflict has been an important one. In more recent research on planning as a communicative process, the focus of interest has been on how conflicts of interest are handled and eventually resolved during the planning process. More important in this contest is the requirement of elevating the conflicts in order to increase the robustness in the planning process (Forester, 1989)

According to practice in the discontinued national physical planning, national-government agencies prepared sector reviews of the national interest as they saw it; then revised them following consultation with national agencies, county administration and municipalities. The Natural Resources Act now assumes this practice will be resumed in preparing structure plans, but national and local planning agencies seem to differ over what the consultations should be about. The issues in question are whether the national agencies can correctly define the national interest and how the national interest so defined should be handled in preparing the structure plans. Municipalities seem uncertain whether national agencies' definitions of national interests are formal declarations or open to negotiation, besides feeling some further specification is needed as to what tangible damage to national interests implies (Boverket, 1992).

This uncertainty is reflected in the planning process. As many as 93% of the plans fail to analyze national interests in a local context and that many of them fail, by extension, to present at all satisfactorily potential conflicts between national and local interests and, more specifically, between conservation and development. As to these general and specific areas of conflict, an overwhelming majority of the plans - 78% and 90% respectively - include no analysis of either of them.

The questionnaire survey, however, revealed a somewhat different picture: 55% of its respondents contended that national interests were presented in relation to local interests and that conflicts between different interests were explicitly presented. Three types of conflicts had a prominent place: (1) between environmental protection and road building,

(2) exploitation and conservation of recreational and natural facilities and (3) national and local interests.

According to the respondents the reason why conflicts were not apparent in the plan document is that municipalities regard plans as the results of resolving conflicts.

The in-depth interviews modify this picture considerably, for they revealed a variety of opinions.

Of the planners and politicians we interviewed, one group felt that municipalities accepted national interests unquestioningly, and that little was called in question because structure planning on the whole was an inventory work; and that much of the process was devoted to gathering data, drawing maps, etc. Some thought that the political aspects were not greatly emphasized. Some planners found they had too little time to discuss the possible implications of the national interests, and that it would have taken much more time to show that local public interests were "worth more" than the national interests.

Another group felt that the designation of so many areas as being of national interest spread alarm in the planning process and laid a dead hand over development. During the process, however, the realization gradually dawned that local planning authorities could actually influence national interests, and a different appreciation of them began to grow.

Yet another group of planners felt that there was a maze of conflicting interests: national as against local, public as against private. Structure planning was too "overall" to deal with these and eventual conflicts between them more explicitly. After all the structure planning is essentially "indicative" in nature. There was hardly any "need to go into this matter any further" than just to show their presence.

Democratic support for planning

Consulting members of the public in the municipality and soliciting their comments was a compulsory part of planning. The implication of this legislative requirement was that public input should be included in the final draft of the plan. The means generally employed were information meetings, public consultations in conjunction with the public exhibition of the structure plan and written comments invited from various associations. But many of them were found inadequate both by members of the public and the planning authorities. Structure plans are more general in their contents, nor is it always easy to identify how different proposals would affect different parts of a community. Municipalities seem to have failed to present their plans in a way that interests the general public.

Accessibility to the contents of a plan is a central issue here, and our evaluation showed that only 11% of the plans we examined provided reasonably good accessibility, by which we mean that people without any specific qualification can read the parts of a plan they are interested in without much effort. One plan in three, however, is difficult for the average person to understand. Of the 29 plans we judged to have good accessibility, 21 belonged to categories four and five, 7 to category three, 1 in category two and none in category one.

The questionnaire survey confirms this picture of public participation: 83% of our respondents felt comments by members of the public had little or very little impact on the final draft of the plan; only 2% felt that public input played a large role in revising plans. Municipal consulting was primarily with other public agencies and the local business community. As many as 66% of respondents said that contacts with voluntary associations, popular movements and other representative groups were poor, very poor or non-existent. Three of four respondents felt that plans could be made more accessible by reducing the volume of legal and technical information in plan documents.

The general picture to emerge is that very small municipalities find it somewhat easier to make adequate arrangements for public participation than larger municipalities. Moreover, municipalities committed to special goals e g establishing an eco-community or restructuring their social-welfare system have been able to arouse public interest more than municipalities which could not isolate such specific issues.

Our interviews modulate the issue of democratic support. According to some of our subjects, Swedish representative democracy is "stretched too far" in the sense that responsibility for making plans is entrusted to only a few members of a municipal executive board. (Many municipalities appoint special groups of municipal councillors for this purpose). The involvement of politicians in plan-making decreases as one moves from the executive board to the municipal council and the political parties. According to one respondent, debate in the municipal council prior to the approval of the plan is often "quite superficial". We find it strange that although both the planning and nature conservancy acts entrust responsibility for planning and environmental protection to local government, local-government politicians have not yet realized how to "make use" of this new situation.

Another issue which our interview subjects took up was the question of who had responsibility for engaging members of the public in plan-making. Those who supposed political parties should do this implied that planners hardly have any "qualifications" to do such a job. It is a question of resources as well as of planning tradition. Planners are accustomed to "working with organized interests", including political parties, trade unions, chambers of commerce, which ensures "an effective and prompt" planning process. A broad consultation process would require a lot of resources and make the whole process "sluggish". Those who claimed public involvement would require an active contribution by planners meant that after all planners have the "necessary knowledge" about the various aspects of planning and, as "public servants", their "obligations are not only towards politicians but also towards the public at large".

Despite the differences in opinions that emerged from our interviews, there was a general agreement that future structure planning that would emphasize "sustainable development" would require a completely different approach to public involvement. Four major sets of issues emerged for this purpose.

Municipalities could make different organizational approaches to public participation: (1) Direct meetings and public discussions, despite a danger of "one-way" communication; (2) Actively engage associations and clubs (sports, temperance, immigrants, ornithologists, greens etc.), rather than merely let them function as "comment groups". (3) Adult evening classes and other extra mural educational activities, which are two highly characteristically Swedish leisure pursuits. (4) Use political parties and their networks, although a drawback is that few people are active in them.

Public participation in planning processes involves cognitive issues. For a municipality, a cognitive approach could proceed "step-by-step", with some simple maps and a few controversial questions being made public in a process that is repeated regularly throughout the planning process. Another is to engage an advertising agency to produce a lavish folder and mount a publicity campaign. A third is to apply Faludi's "planning doctrines" (Faludi and van der Valk, 1994) and isolate specific issues and arouse public interest or, alternatively, by isolating several issues, to exploit a "single symbol issue" to stir up public interest.

Suggestions of five appropriate stages for public participation were put forward: when structuring problems; during discussions on goals and policy measures; during consultations when a preliminary draft is presented; when the final plan proposal is exhibited publicly; or continuously from start to finish of the process.

Among the resources available for public participation, planners named political parties, associations and clubs (as above), municipal planners and other local government officials.

Knowledge and values

Notably unlike the regulations that it superseded, the Planning and Building Act leaves municipalities to draw up their own plans, but even if the Swedish National Board of Housing, Building and Planning no longer issues detailed instructions, the old blue-print and rational-planning methods seem still to exert their influence. Of the respondents, 51% expressed a belief in rational-planning methods but 49% had little confidence or less in them.

We have had difficulty in describing municipalities' procedures for preparing their structure plans. About 11% of the plans contained some account of the planning procedures that produced them, but most accounts were brief and offered no real insight into how the planners had worked. The belief expressed by 51% of planners in rational-planning methods is hard to interpret in the light of the opinion of over 66% of the planners that planning was considerably characterized by so-called incrementalism.

In their work, the planners used progressively less quantitative information derived from projections and surveys, and gradually replaced and complemented it with values and personal experience. However, most of them still had great confidence in quantitative methods.

Our interviews showed that there are several "problems" related to balancing knowledge and values.

Planning legislation relies on centrally derived "norms and requirements" that are not based on an appreciation of the planning process as it actually takes place locally. This makes it hard for planners to develop a plan that politicians and the public can easily understand. Even when attempts were made to simplify the text of plans, lawyers and land-surveyors found simplification "unattractive", and objected (see also Gustafsson, 1987, for similar observations).

Structure plans extensively used earlier plans, of which many had been prepared during "the heyday of rational planning". It was often difficult to "complement" this knowledge with values. Together with planners' poor ability to formulate issues so that the public takes notice, this puts a heavy bias on "processed knowledge".

Structure planning was rather too "programmatic" in character, included "too much stock taking" and a "lot of description" and put forward "ideological viewpoints" far too rarely. To analyze and explain conflicts between various types of public and private interests requires a "great portion of values". Planners were so involved in the "crude work" of drawing maps and preparing the plan text that they could never "to get down to" interpreting knowledge and values in this context.

8.6. Concluding reflections

This paper presents a methodology for assessing the political dimension of planning process and its application to the structure planning process in Sweden.

The methodology is based on a simple framework that includes some major aspects of the politics of planning. It is flexible, in that these aspects and the variables attached to them can be varied. In this paper the political assessment has been limited to the three aspects of the legitimacy of planning as expressed through the pursuit of public interest; democratic support and public participation; and the integration of knowledge and values during the planning process.

The paper raises several questions at a theoretical level. Is such an assessment useful? If it is, how should one make it? If a theory and methodology of assessing the planning process is necessary, then it is essential to examine further variables. In finding them and forming an assessment matrix for each, both planning theory and feedback from planning practice must be used. Assessment of planning process is an intellectual and practical challenge. Both post-modern and critical theory emphasize the communicative nature of the planning process. An important contribution towards a communicative theory of planning is to develop a set of criteria for evaluating the planning process.

At an empirical level, the assessment throws light on some central issues in planning process.

The pursuit of the public interest is fundamental to all public planning efforts, and all the central issues with regards to it - namely opportunities to protect it, planners' ability to ward off threats to it, and their response towards balancing national and local interests, etc. - have received inadequate consideration in the planning process. This is indeed very serious not only because it raises doubts as to the legitimacy of planning efforts but also because in the current devolution of power from the central to local governments, the latter have been given greater share of responsibility in such important areas like environmental policy. This requires proper attention to issues related to various types of interests.

The lack of public involvement in planning, and thus democratic support for planning, is an important challenge to planners in Sweden, where a tendency to replace politics by administration and expertise has been strong. This, together with over-much faith in representative democracy, explains why many ideas about public participation have never been put into practice.

In order to consider how knowledge and values relate, and how this relationship affects the structure planning process, we need to have access to the political process which led to the final approval of the structure plans. Without it, we tried to find out the role of scientific or processed knowledge as generated in long-term projections and quantitative examinations in relation to knowledge related to people's everyday life. Municipalities lack what is needed to co-ordinate and balance these two types of knowledge. In such circumstances it is not surprising that only few municipalities examined the interplay between social, economic and environmental goals of planning.

There are extenuating circumstances for the unsatisfactory state of structure planning in Sweden, which was introduced as the whole central-planning system in Eastern Europe was collapsing. The neo-liberal critique especially as espoused by thatcherism turned out to be very powerful. At a time when the whole idea behind anticipatory planning was under challenge, it was not surprising to find the low level of ambition as reflected in the planners' conception of the status of the plan as well as in the political involvement in the plan-making exercise.

References

Benveniste, G. (1989) Mastering the Politics of Planning. Jossey-Bass, San Francisco.

Bohm, K. (1985) Med-och Motborgare i Stadsplanering, Liber, Stockholm.

Boverket (1992) Erfarenheter av översiktsplanarbetet, The Swedish National Board of Housing, Building and Planning, Karlskrona.

Catanese, A. (1974) Planning and Politics: Impossible Dreams, Sage, New York.

Davidoff, P. (1965) "Advocacy and pluralism in planning", Journal of the American Institute of Planners 31, 331-338.

Eckstein, A. (ed.) (1971) Comparison of Economic Systems, University of California Press, Berkeley.

Faludi, A. and van der Valk, A. (1994) Rule and Order: Dutch Planning Doctrine in the Twentieth Century, Kluwer, Dordrecht.

Floderus, Å. (1986) "Vad hände med det planmässiga samhällsbyggandet?", in Floderus, Å. (ed.) Planering med Nya Lagar, Swedish Council for Building Research, Stockholm.

Forester, J. (1989) Planning in the Face of Power, University of California Press, Berkeley.

Friedmann, J. (1987) Planning in the Public Domain: From Knowledge to Action, Princeton University Press, Princeton.

Gustafsson, G. (1987) Decentralisering av Politisk Makt, Carlssons, Stockholm.

Healey, P. (1989) "Planning for the 1990s", Working Paper no.7, Department of Town and Country Planning, University of Newcastle Upon Tyne.

Healey, P. (1992) "The communicative work of development plans", In Khakee, A. and Eckerberg, K. (eds) cit.

Howe, E. and Kaufman, J. (1981) "The values of contemporary American planners", Journal of the American Planning Association 47, 266-278.

Khakee, A. (1978) "Municipal planning in a mixed economy", Environment and Planning A 10, 437-450.

Khakee, A. (1994) "A methodology for assessing structure planning process", Environment and Planning B

Khakee, A. and Eckerberg, K. (eds.) (1993) Process and Policy Evaluation in Structure Planning, Swedish Council for Building Research, Stockholm.

Khakee, A. and Dahlgren, L. (1990) "Ethics and values of Swedish planners: A replication and comparison with an American study", Scandinavian Housing & Planning Research 7, 65-81.

Lutz, V. (1969) Central Planning for the Market Economy. An Analysis of the French Theory and Experience, Longmans & Green, London.

Masser, I. (ed.) (1983) Evaluating Urban Planning Efforts, Gower, Aldershot.

Miller, T. and Kraushaar, B. (1979) "The emergence of participatory policies for community development: Anglo-American experiences and their influence on Sweden", Acta Sociologica 22, 111-133.

Ödmann, E. (1992) Planlagstiftningen och Välfärden, SIB, Gävle.

Thornley, A. (1991) Urban Planning Under Thatcherism, Routledge, London.

Vasu, M.L. (1979) Politics and Planning. A National Study of American Planners, The University of North Carolina Press, Chapel Hill.

Abdul Khakee
Statsvetenskpliga istitutionen
Umeå Universitet
S-901 87 Umea
Sweden

9 EVALUATION OF QUALITIES IN SPATIAL PLANNING PROCESSES

Riccardo Roscelli

9.1 Plan as a process of change

If a project or a plan is seen to be a process of change and action able to modify the preexisting state, then the process is the mix of quantitative (amounts) and qualitative (types) elements as well as issues being described according to a multi-dimensional approach.
A first set of issues relates to the following: I) measuring methods, II) how project or plan feasibility is ascertained, III) identification of possible conflicts that the project may highlight, and IV) defining the time and space horizon that may at times radically change the value judgments themselves.
The second set of issues pertains to the notion of quality itself. I shall be dealing with this issue further on in the text. For the time being I shall merely stress its relative rather than absolute features.
The planning process is often thought to coincide with the papers and drawings that are in fact only the image of the plan, its physical and technical structure. The rules determining plan compatibility when moving from the design to the implementation phase and subsequently to its use are not so clearly understood. At this stage, features such as economic compatibility, appreciation (or depreciation) processes and standard (or quality) assessment emerge and have to be dealt with. Otherwise, the issues that come to light when the plan has to be implemented change the previous balance, whether or not the plan refers to a building, a facility or a town plan but all the more when it concerns a process of changing land use. In this framework, the value judgments on the actual or presumed qualities that the plan refers to must be considered carefully: mainly personal or subjective judgements may not successfully stand the test of time, nor can they easily be measured according to clear and unambiguous rules.
Let me offer a few examples, such as that of measurements. I am using this example only for the purpose of agreeing on the words and definitions, at least for the more straightforward cases. A plan comprises a number of items that can be measured using quantitative units from physics: distances measured in linear metres, volumes in cubic

D. Borri et al. (eds.), Evaluating Theory-Practice and Urban-Rural Interplay in Planning, 117–122.

metres, surfaces in square metres, and so on. In the event of elements that are comparable among themselves, and can be compared to the same unit or measurement, albeit not necessarily with a very high degree of accuracy, then an ordinal scale of numbers can be used. The scale may consist of generic elements, such as 1st, 2nd, 3rd, 4th and so on, without stating the distance between the first and second, the second and third and so on. Alternatively, the distance or gap can be measured with satisfactory accuracy by using physical units of measurement, so as to establish the distance between the first and second element is a percent ratio.

However, one is often faced with the need to assess quality: quite a different ball game, since we often have to resort to nominal value scales and to languages that differ substantially from the formalised mathematical definitions that can be reduced to algorithms of varying complexity. In this case one might say the plan is more beautiful or not as good as another. Although this judgment may be interesting and effective, it is not so definite nor clear. For instance, when I am lecturing I offer my students very simple examples, e.g.: when the Colosseum was built it was unquestionably a symbolically relevant building. It then became a historical milestone of Rome's Imperial power, as well as the place where Christians were persecuted. Despite that, it was used in the Middle Ages as a quarry, and its stones were taken to build less precious edifices with a very great use value, and its symbolic value was lost. Currently it virtually epitomises Rome itself and is a listed building. How can we judge the value of this building? In more general terms, could one suggest the notion of a spatial building able to embody quality evaluation in a framework of land use change over time? [1]

In our subject, we appraise and evaluate to select one or more suggestions. Priorities have to be set when economic resources are scarce; appropriate uses have to be identified when existing tangible architectural and heritage features have to be preserved; economic and quantitative values have to be listed according to a hierarchy, and mainly cultural symbolic and cultural values have to be considered.

These problems are always found in built up areas, and also in natural environments: hence the need to identify the values at stake openly and clearly, to specify the criteria or parameters used in selecting and choosing the plan. On the whole the above are also used to define project viability.

I am now going to offer examples which I hope will illustrate the degree of complexity we have to deal with: the consumption of a resource (or ought I say of a series of resources) or the downgrading of a standard, be it due to decline, or to a diminished cultural and symbolic value, may be taken as the amount of environment-capital available to present and future generations, the capital to satisfy the demand for goods and services. Hence the definition I gave of the 'spatial monument'. In the event of a decreasing environment-capital, this may be perceived as a net loss. If so, one must be able to quantify it, especially if they are unique goods that cannot be re-produced with strong interactions in the area; let us presume they generate long-term effects whose impact exceeds the life of the goods themselves. Furthermore, they also affect the value judgment when and as we evaluate them. Such features clearly do not lead to easy solutions.

There is a wealth of scientific literature on the topic. Two separate positions emerge, along side a number of intermediate and compromise solutions: the first maintains it is always necessary and possible to evaluate the economic value since listing and

conservation are *per se* productive activities; the second stresses the meaningless of using an economic rationale in a system of values that does not consider them and which is furthermore unable to express values in monetary measurable units[2].
One might fruitfully compare the results of the two approaches in theory and in practice, so as to understand their potentials and limits when applied to a number of unresolved practical and theoretical issues.
In view of the decreasing availability of environmental and territorial resources, issues of their conservation have become increasingly important, and an economic approach capable of understanding appreciation (or depreciation) over time has become more interesting and effective.
Increasingly, this kind of 'qualitative' resource, such as our cultural architectural and environmental heritage, determine the values underlying economic or symbolic values. Furthermore, such goods are characteristically scarce, unique not reproducible and often not usable *per se* since they are part of a more general context. Furthermore, these resources can be assessed in terms of future users, also in view of their inherent ability to preserve 'quality'

9.2 Evaluation procedures

This is why unambiguous procedures must be laid down when outlining evaluation strategies. I am not thinking of technical difficulties, but rather of the complexity of the process involved. For instance one cannot refer to a market and to specific prices, to demand and supply levels, or to specific and distinct aims: we have to move in terms of forecasts. The boundary between logic and field of action in terms of economic viability and estimated judgment values may sometimes overlap, and are often left undefined in current scientific debate.
The research group I belong to has been working on quantitative-qualitative evaluation, hierarchical analysis, discordance and concordance analysis, regime analysis and so on. Other limited economic analyses were tested and are currently being tried out in experimental studies and simulations, such as Hedonic prices and Clawson's travel costs, option, existence, complex social and contingent evaluation values. In the case in point, these philosophies and procedures are quite sophisticated, theoretically interesting, but have a number or grey or uncertain areas, as clearly happens when dealing with matters of this nature[3].
I am deliberately keeping to the margins of the debate about what money can and cannot measure, a topic that appeals to both economists and valuers, and an area where new certainties are followed by disappointments. I would like to deal with minor but central issues.
I will omit the description of how direct and overt benefits, as in the case of setting user fees, are measured. Widely accepted methods and of other tools used in the cost/benefit method (although in a situation of greater uncertainty) to evaluate intangible effects or goods without a market, or with potential but not actual markets include the following: i) establishing other prices, for instance using prices for comparable goods elsewhere; ii) the willingness to pay, that is to say the price consumers would be prepared to pay if the goods were on the market, on the basis of market surveys of the potential market; iii)

behavioural analysis, already described as Clawson's method (Clawson, 1959) based on the costs people bear, on the access they have to goods, or the use they make of such or comparable resources.

None of these appraisal methods is perfect: being fundamentally based on forecasts, they all have shortcomings. However accurate the forecasts, they are highly unpredictable and therefore risky; furthermore, the markets and socio-economic contexts they operate in are also mixed and results are not always reliable.

When trying to identify the so-called shadow prices - and some indirect benefits such as employment or income multipliers - the input-output analysis (IOA) may prove to be a useful tool in measuring effects for which no market exists at the time of appraisal. IOA can be applied to inter-sectorial tables of economy. I believe this method could be convincingly used to identify the existence value of existing environmental resources or facilities, and of features of cultural, architectural and landscape heritage.

Furthermore, IOA also enables us to identify the relationships between and among sectors of industry, and consequently the impact of a given resource on the entire economic system as well as the micro-effects of introducing a given technology or its subsequent ripple effect in relation to demands and to consumer needs[4].

Albeit only as a method, IOA may help us find a solution to appraise intangible aspects, that have a potential to damage the environment and which need to be evaluated, whether or not action is taken.

In other words, Klassen's approach to the "shadow plan" (Klassen, 1980) could be experimented along these lines. The plan could prevent or compensate for damage to the environment by offering appropriate economic or monetary redress. This would yield two immediate advantages: reducing the problem and quantifying or measuring what is otherwise incommensurable and making certain that the damage to the environment will be compensated for even if only relatively.

The main problem in this process is related to setting up the simulation.

Leontief proved how the IO model can be modified to appraise environmental variables by treating the activities that affect the environment as sectors of the system. They will be considered equivalent to other economic activities, and special attention will be devoted to how the aforementioned relationship in turn impacts on the economics of the supply system. As a result, the flows of damage could be expressed in terms of the resources required to eliminate it, as well as defining the economic activity able to counter-balance its negative effects (Leontief, 1970).

In this way it would be quite simple to measure (or quantify) the environmental value by default as the equivalent of the damage it would have suffered had the action not been taken.

Once this problem has been solved, the remaining evaluation could be carried out with more established methods drawn from the Contingent Evaluation (and subsequent modifications). Adjustments would be needed to suit the specific infrastructural features that are going to be used, while taking the indirect and induced economic effects into due consideration.

In practical terms, it implies an infinite time horizon, legislative protection for all environmental resources, a reduced rate of discount, and adequate price systems, as mentioned in the case of the shadow prices and the so-called availability to pay.

9.3 Concluding remarks

I have tried to illustrate how i) it is impossible objectively to measure (quantify) some of the ingredients of value; ii) other ingredients belong to the category of unmeasurable goods and can therefore only be indicated pointing out the weight and influence they have in that case; and iii) to list projects according to preference or indicate alternative routes.
To reach the discrete solution to problems as complex as these, patient, careful interdisciplinary work is needed, which is what we are attempting in the course of the present debate.
I believe that in the present phase one must try and cover at least part of the enormous gap between, on the one side, the practice and theory, and, on the other, epistemology. It is a no man's land that can be conquered by neither side, but which is a fascinating experimental and innovative ground for both. In the present uncertain and rapidly changing climate the evaluation of plan feasibility is very important in the context of the plan-market sphere: a multi-disciplinary approach is required for these issues that lay in-between practice and theory.

References

Clawson, M. (1959) "Methods for measuring the demand and the value of outdoor recreations", in Resource for the Future, 10, reprint.

Fusco Girard, L. (1989) Risorse Architettoniche e Culturali: Valutazioni e Strategie di Conservazione, Franco Angeli, Milano.

Grillenzoni, M. and G. Grittani (1990) Estimo, Edagricole, Bologna.

Grittani, G. and R. Roscelli (1991) "Economic evaluation models about water resources", in Medit 1/2.

Klaassen, L.H. (1980) De Onderzoeker en de Politus, Beleidsanalyse

Leontief, W. (1970) "Environmental repercussion and the economic structure: an Input-output approach", Review of Economic and Statistics 52.

Lichfield, N. et al.(1990) "Rapporto sull'analisi costi benefici per il patrimonio culturale costruito", in Restauro 111/112.

Pearce, D.W. and K.R. Turner (1991) Economia delle Risorse Naturali e dell'Ambiente, Il Mulino, Bologna.

Posocco, F. (1991) "Insediamento e pianificazione nella storia del territorio", in Proceedings of the Meeting on "Trasformazioni dell'uso del suolo e conseguenze sulla rete idrica del Veneto", Venezia.

Roscelli, R. (1984) Le Principali Variabili dell'Attività Edilizia, Celid, Torino.

Roscelli, R. (1985) Misurare nell'Incertezza, Celid, Torino.

Roy, B. (1985) Methodologie Multicritére d'Aide a' la Decision, Economica, Paris.

Voogd, H. (1983) Multicriteria Evaluation for Urban and Regional Planning, Pion, London.

Notes

[1] For a re-definition of the term, see F. Posocco (1991)

[2] Fusco Girard (1989), Pearce and Turner (1991), Lichfield et al.(1990), Roscelli (1990), Voogd (1983), Roy (1985).

[3] See, for example Chapter 5.5 on quantitative-qualitative evaluation, edited by Roscelli R in Grillenzoni and Grittani (1990).

[4] Roscelli (1984), Grittani and Roscelli (1991).

Riccardo Roscelli
Dipartimento Casa Città
Politecnico di Torino
Viale Mattioli 39
Torino
Italy

10 PROBLEMS OF URBAN LAND-USE AND TRANSPORTATION PLANNING: COGNITION AND EVALUATION MODELS

Angela Barbanente
Dino Borri
Valeria Monno

10.1 Introduction

The paper[1] deals with the crucial issue of reconciling transport policies and sustainable development at the urban level.

There is little doubt that accepting this sort of challenge would imply the strategic coordination of environmental, land-use and transport policies that are far from being achieved, in either theory or practice. In this perspective, our paper focuses on the following issues.

Firstly, we try to clarify our point of view on the use of the sustainability concept, to investigate the principles underlying it and what it implies for defining transport policies.

Secondly, we highlight some problems deriving from current approaches to transport/land-use analysis and planning, both in theory and in practice, with particular reference to recent trends in making transport policy in Italy.

Thirdly, we propose a case study in order to look thoroughly at experts' cognitive and evaluation models as a preliminary to proposing different ways of setting transportation problems and, consequently, of defining integrated transport/land-use strategies for sustainable urban development.

10.2 Sustainability and transport planning

When the environmental question became acute, the problem of transport planning seemed to be critical for energy consumption and for limiting the greenhouse effect. At the same time the overall impact of urbanization has clarified the strong connections between environment degradation, transport and land-use.

In this context, sustainability, which is a necessary reference to form effective knowledge bases on which to start environmentally compatible urbanization processes (Maciocco, 1991), represents a reading that could contribute to explaining as yet not clearly unexplained connections between transport and land use (Hart, 1992).

D. Borri et al. (eds.), Evaluating Theory-Practice and Urban-Rural Interplay in Planning, 123–139.

Sustainability. Even within the perspective of sustainability, various approaches and courses of action may be proposed as possible ways to manage and control problems linked to transport and land-use.

Sustainability, a vague and ambiguous concept, represents a "story line" (Hayer, 1992), rather than a definite alternative development theory that draws together different points of view expressing a plurality of possible relations between economy and ecology. Consequently the real development perspectives linked to it depend on the number of interpretations so far made of it (Archibugi, Nijkamp, 1989) that differ in their treatment of the three key elements that underlie it:

- equity: the achievement of widespread social justice in the distribution of and accessibility to resources in both space and time;
- environment: acknowledgement of nature's rights and values;
- development: economic development capable of guaranteeing both the quality and quantity of natural resources.

In a co-evolutionary perspective the possibility of orienting environmental policies towards a sustainable model depends on beginning by acknowledging nature's right as not negotiable, and by basing policy more on integrating ecology and the economy rather than only accepting "market" solutions that may seem partial, although necessarily supportive.

Integration should not be thought of as a simple linear superimposition, for it should imply that the environment resembles a complex system, in which the risk of irreversible, uncertain, unknown, unpredictable phenomena and actions may occur.

Planning and sustainability. The conflicting differences between the perspectives of natural and artificial systems explain the need for ecologic and economic integrated models that make it possible to plan acceptable social systems for the environment. As a matter of fact, planning can also play a very important role in achieving sustainable development (Blowers, 1993) because it represents a way round the widespread "limited" approach to problems of environmental management caused by partial legislative action by local government (Van der Valk, 1990). Thus the environment, entering the planning process more as an evaluation parameter than as a value, can radically change not only public decision-making processes, but also the microdecisions (Borri, 1993), that characterize all planning activity.

Sustainability introduces some crucial factors of necessity into the planning process:

- it puts sectoral and local interventions within a general, long-term programme. In a context of international competition between cities and regions, it is necessary to recognize challenges and opportunities, the chances to combine the need for local development with global sustainability strategies ("Think globally and act locally" is a must);
- considering the dynamic character of phenomena and systems for intervention, it acknowledges that the future is unpredictable. By making and pursuing strategic choices for short-to-medium and long-term periods, it is to some extent possible to compensate for an unforeseeable future by making at least the goals for public and private action easily predictable;
- it links decisions to: (i) a process of identification of alternatives based on current environmental dynamics and on possible future circumstances; (ii) negotiation processes

necessary because environmental problems cause conflicts and because the process needs to include a possibility of modifying action and reorienting the decision-making processes;

- it provides a means of monitoring the course of the action in the perspective of concrete action (Maciocco, 1993) based on the principle of caution.

The central role played by information is very clear and necessary because of the complex, multidimensional character of the phenomena involved and because of the need to follow a non sectoral approach. Sustainability requires a multidimensional, cohesive and strategic way of thinking (Archibugi, Njikamp, 1989).

Analysis of the environment plays a double role: it is exploratory in aiming at describing environmental issues, and it is operative as a tool for determining a general outlook of environmental compatibility, of resources and non-negotiable meanings-values. Such analysis that starts with the dynamics of current processes can be a basis for building a potential future and can also identify different sustainable patterns and assist a choice of acceptable ones (Braat, 1991).

Transport/land use. By integrating transport policies with the configurations of spatial settlements and the specific environmental dynamics of the relevant area in question, analyses of how transportation and land use relate to each other should generate urban and regional development scenarios that would reduce pollution phenomena and the depletion and degradation of environmental resources.

At present, the concept of sustainability is remote from an understanding of the mechanisms being used for land organization increasingly characterized by mobility (Vittadini, 1992). As to transportation, a series of urgent measures include:

- using alternative technologies, clean petrol and giving incentives to pedestrians and cyclists;
- reinforcing electrical public-transport facilities and multimodal systems;
- reducing car use by means of taxation measures or others based on polluter-payer principles;
- recovery of existing road infrastructures;
- priority for transporting freight by rail.

It is more difficult to reconcile these measures with the dynamics of local and global territorial developments and to intervene in the mechanism which is at the roots of mobility.

Information technologies/transport. Moreover the ability of new information technologies to change traditional communications makes transport policies that ignore traditional systems together with immaterial ones increasingly unsuitable. But to conceptualize the relationships between land use and new information technologies seems debatable and problematic. In fact the use of new technologies and all this implies for territorial organization still seems conflicting and divisive. While they may be able to modify relations in ways of communications, they are less able to modify the spatial distribution of the activities (Scandurra et al., 1993). In any case, they do not establish equipotentiality (Bertuglia, 1993) as regards possible global v. local conflicts, and they do not reduce spatial mobility.

Models and problems about knowledge. In this context reference to sustainability can identify some crucial problems, which would contribute to an understanding of transport/land use relations and to a formation of knowledge bases able to interpret them.

A first step is grasping the essence of problems of relations between transport and land use, and reformulating them in terms of the goals and/or values of sustainability. In fact, introducing equity and environmental-preservation criteria shows that efficiency is inadequate as the sole criterion for resolving a transport/land use problem, while it remains necessary to resolve the controversy of whether land-use policies can to a greater or lesser degree reduce energy consumption and pollution phenomena (Wegener, 1993). With a view to sustainability, to plan transport must be done in an overall integrated strategic way that directly considers land-use patterns based on energy cycles and ecosystem resources.

A second aspect also derives from defining and recognizing the key variables of problems of transport as against land use: the adequacy of methods for representing environmental dynamics and whether they can be used in dealing with such problems. The potentiality of traditional patterns in building scenarios, although they are considered unable to cope with highly complex problems, point the way to the reframing of classical concepts and principles, such as accessibility and gravitation, in relation to increasingly complex and evolving settlement dynamics and to rapid changes in traditional communications caused by new information technologies.

Phenomena of mobility are becoming less systematic in character and indeed seem increasingly subjective, which makes aggregated models less interesting than the disaggregated ones that can underline the relevance of individual choice mechanisms and the related links with the activity system.

Traditional models wholly lack any environmental dimension. To remedy this, geographical pictures are urgently needed of the complex relationships between environmental patterns and spatial organization, which multidisciplinary analytical methods and the treatment of qualitative and qualitative information can provide.

If no analytical operative tools are available that can clarify the complexity of environmental dynamics, studying the relations between the transport system, seen as one material and immaterial network, land use, as a whole characterized by the relations between environmental patterns and settlement typologies, as well as mobility can contribute to explain, recomposing different patterns of knowledge in building scenarios, how urban form, on one hand, and travel behaviour, on the other, are able to mitigate the harmful effects of transport system on the environment.

In connection with possible policies for material and immaterial transport it is necessary to analyze, opposing prospects of territorial development linked

(i) to models of compact cities - with low energy consumption and low harmful emissions (Kenworthyn, Newman, 1989), but which, from on environmental point of view, seem to intensify the conflict around green areas and unoccupied urban lands, and that do not necessarily imply mobility reduction;

(ii) to models of network development - which, on the other hand, might seem to be directed towards a further consumption of resources and increase in pollution, while, on the other, they offer the possibility to rehabilitate the building heritage and infrastructures and reduce travelling distances.

This may contribute to selecting relevant variables for sustainability to be empirically assessed applying the monitoring of action to that end. At the same time the empirical evaluation of consolidated theories may reveal their possible weakness and consequently the possible way to correct them in order to make the theories fitter for conditions as they stand (Giuliano, 1988).
The goal is therefore also to find analysis models and investigate their potentiality for simulating and therefore for setting up possible sustainable scenarios.

10.3 Land use and transportation planning: seeking an environment-oriented integration

Since the publication, in 1954, of the famous work by Mitchell and Rapkin, the existing reciprocal, and very close, relationship between land-uses (related to corresponding activity systems) and mobility flows has become very clear. Though, in the version of that proposal that mostly influenced further work, mainly in transport planning practice, only one direction has been grasped, namely, traffic and transportation as dependent variables of a certain spatial distribution of activities [2].
The typical traffic engineering approach to transport policy is still well-established. This, within the dialectic relationships linking transport supply and demand, privileges a definition of supply as a function of demand and, as we can see has happened very frequently during the last fifty years, given the increasing demand for mobility, thus sets up the need to build new transport infrastructures, mainly new roads, without promoting different transport modes (the so-called green modes) together with appropriate traffic calming measures, in an attempt to meet the accessibility demand differently.
Furthermore, that kind of approach tends largely to ignore the responsibilities of mobility policies in determining mobility patterns and spatial transformations, and the effects, not rarely perverse, of transport policies essentially based on increase in infrastructures supply: it is clear, on the contrary, that increased investments in new roads, new parking areas, etc., themselves generate new traffic flows, and while they resolve one problem, they generate new ones.
Breaking down this vicious circle probably implies a reconsideration of the problem with respect to conventional views, investigating the role that has really been played by transport policies in the national economic system, on the one hand, and the actual outcomes of investments in the transport sector in the light of the spatial changes, both from socio-economic and environmental viewpoints, on the other.

Roles of transport policies and inherent conflicts: trends in Italy

First, we need to ask ourselves whether transport policies have really been dominated by supply, at least in a stage of Italian economic development. In fact, there is little doubt but that post-war infrastructure policy in Italy assumed roles that go very much beyond their traditional espoused goals: we must think in particular of the role, once more unfavourable, that economic trends played through investments in the transport sector, mainly in motorway works; and also of the support given to the car industry by infrastructure policies [3].

Furthermore, a policy based on huge investments in transport networks, particularly in motorway networks, received widespread support in Italy in the last post-war period, because of convincing arguments in its favour: we refer to the supposed benefits it promised for economic gaps at regional and sub-regional levels. In the first case, most experts and decision makers thought that, by promoting connections between underdeveloped and developed areas in Italy, it would be possible to attract further industrial investment to the areas that were to benefit from infrastructures; in the second case, they thought to foster the decentralization of productive activities, in order to restrict growth and congestion in central areas and to develop peripheral areas.

As a matter of fact, the motorway policy was particularly to favour developed areas, as it mainly enlarged the markets of big industrial companies in Northern Italy (Secchi, 1973), and permitted sub-regionally more dispersed patterns of urban development together with a rapid growth in car ownership.

Furthermore, energy consumption and pollution problems, mainly originating in urban life, require a new strategic approach to urban land-use and transport planning, and a new focus on urban form design, also including the environmental perspective (Breheny, 1992), and this appears a difficult task in the light of recent changing trends in the spatial-temporal dynamics of post-modern society (Harvey, 1989; Castells, 1989).

Particularly within large urban areas, traffic congestion, and consequent environmental pollution, which tend to occupy widening spatial ambits and longer time-lags, makes it increasingly difficult not only to perceive solutions to these problem, but even to define them and to understand their causes (Zambrini, 1985). On the one hand, processes of selectively centralizing tertiary functions to metropolitan core areas (often as part of central government policies that rely on the greater competitive potential of central areas), increases theoretical mobility and real parking demands in areas that absolutely cannot provide them; on the other hand, the persisting tendency to decentralize economic activities that cannot compete in central areas (because of increasing real-estate prices, dependence on locational accessibility, and needs for large amounts of space) and of settlement spread for housing and consumer services and recreational facilities, tends to encourage private motorization and to worsen the environment in increasingly wider areas.

Within these trends, achieving the goal of avoiding excessive densities (at given levels, the energy advantages are lost to congestion problems, inefficient energy use and loss of accessibility) and dispersed patterns of urban development, is a difficult challenge for land-use planning (Owens, 1992).

When we consider that governments promote innovation to increase local competitiveness, this calls for simultaneous interventions in spatial policy to make these localities environmentally attractive, we can fully understand the contradictory character of the policies that pervade current trends.

In the light of these considerations, it is difficult to accept either that transport policies merely comply with mobility demand by means of sectoral or contingent measures, or that transportation planning and land-use planning develop separately by following their own sectoral logic; according to a familiar Italian procedure, the first takes the second as

given (of course, this is not an Italian peculiarity: for the French case, see Merlin, 1984; for the British case, see Plowden, 1985).
But supposing we are trying to pose the problem of land-use and transport planning differently, we cannot just note the fact; we must investigate its causes. We can then adopt either of at least two related perspectives (or all of them): one focuses on organizational and regulative aspects, the other on current knowledge and evaluation approaches that people (in particular experts) use when addressing land-use and transport problems.

Organizational and regulative aspects

Firstly, we must note that the functional separation of ways of dealing with land-use/transport problems is reflected in administrative organizations in charge of planning/managing them: both regionally and locally, most town-planning and management departments and transport departments are structurally separate and usually operate according to power and consensus logic that obstruct - and in any case do not foster - the integration processes of sectoral policies.
We have also to consider that, at present, local planning has a limited role to play in the definition of integrated land-use/transport policies: Italian transport infrastructures are entirely within the large public works sector that is dominated, even nowadays, by centralized policies, especially as far as resource investments and the definition both of subjects to which they have to be assigned and works typologies are concerned. Moreover, the current increasing concentration of investments benefit big projects and big operators, and consequently favour direct relations between big public agencies and corporations (Government Departments, ANAS, FFSS) and big private business companies, whereas local authorities, when taking decisions in land-use matters, become subordinate to centralized powers, as the decisions bringing about the real implementation of infrastructure plans are taken elsewhere.
Thus, projects are superimposed on the plan, and often contradict it (a procedure permitted by the terms of the relevant laws), with the result that significant spatial transformation processes come from interventions and subjects external to a coherent planning framework laboriously built at local level [4].
The lack of connections between the National General Transport Plan and spatial planning courses of action at regional level; the admitted ineffectiveness of regional transportation plans and traffic areas plans - respectively entrusted to regional and local authorities since 1981 - due to the partial competence of these authorities in the transport sector, as whole key branches lie within the competence of central powers; the lack of a regulative definition of the goals and contents of urban transport plans; the introduction of urban traffic plans, which are defined as short-term intervention plans (their adoption, in 1986, was recommended for municipalities above 50,000 inhabitants, and then, in 1993, was made compulsory for municipalities above 30,000) without any advice as to their relations with town planning; a body of legislation for the realization of parking areas (introduced in 1989) that foresees the possibility of carrying out projects that failing to meet planning prescriptions; all these issues point out not only the problem of the definition of coherence criteria for different planning typologies, and between these and large-scale projects, but most of all that of making the goals and interests of different

agencies and operators explicit, this being a fundamental premise for any prospect of coordinating them.

Furthermore, the above mentioned mechanisms of distributing public-expenditure flows in the transport-infrastructures sector give rise to a system in which the availability of financial funds for certain work induces local municipalities to "set themselves in motion" in order to get the funds. This is not an anomaly. The perverse aspect of this mechanism consists in the fact that local governments project solutions, not to specific local problems, but with reference to the provisions of laws worked out centrally, often without making any evaluation of the putative solutions themselves.

Good examples of such distorted processes are some recent events to do with urban parking areas (National Act 122/1989) and of rapid mass and bound driving transportation systems (National Act 211/1992).

As to parking areas, a legislative financial measure has advanced a limited part of the complex matter of improving traffic conditions in large urban areas, which has self-defeating effects on traffic levels, especially because neither the parking-areas location nor relations between this and other actions in land-use and transport fields is adequately evaluated.

In the second case, ever since the legislation was enacted, a putative solution has been defined in terms of transport mode and technology (particularly innovative transport systems) and without knowing whether this would be justified by specific local problems. Most probably, the choice of technology was not really to solve some local problems, but mainly to further a nationally relevant industrial policy, i.e. supporting innovation in the transport industry that, being a mature branch, had serious unemployment and market problems (Fareri, 1990; Vittadini, 1991).

In this paper we are not going to touch on the different theoretical interpretations of land-use conflicts, as this would go far beyond the scope of our contribution. Nevertheless we wish to note that, also because of its centralized institutional and operational context, transport is one of the sectors rife with inter-organizational conflicts whenever large projects of strategic local development are in prospect. The sector can mobilize large financial resources, making it really hard for the agencies concerned (typically, local regional and municipal bodies, central government departments, state-owned enterprises) to reach agreement.

But also social conflicts arise over obtaining local communities' agreement with respect to infrastructure projects that often have strong environmental impacts. In a sort of vicious circle, this can have repercussions on inter-organizational conflicts, as local governments are under political pressure from local voters.

Thus, the local character of land-use policies - and the essentially local views of opposition groups that can usually not see things in a wider framework (Saunders, 1981) - assumes a much wider dimension. In the light of our foregoing considerations on the aims and the developments in transport policies, exclusively local considerations obstruct an efficient location of projects really aimed at general economic expansion and politically controlled conservation (Plotkin, 1987).

Having in mind this general frame of reference, we find that a routine treatment of these problems, which the following experiments of participant observation well exemplify, contains no trace of conflicts over land-use in contemporary society, at geographical and sectoral levels (local municipalities interests versus regional or national interests, consumers v producers interests), especially when environmental matters are concerned. It may be of some interest to look at some specific case studies to try to work out a way of defining the problem differently.

10.4 Expert cognitive models for managing land-use environment conflicts in transportation problems

As regards the critical issues addressed in Sections 10.2 and 10.3, we decided to carry out an experiment of participant observation. We asked two 'experts' to solve an urban transportation problem embodying some of those issues.

The experiment also aimed at focusing on the potentials of Artificial Intelligence (AI) methods for an automated solution to transportation problems, on which not much literature exists (but see Witulski, 1992).

The participant observation (PO) of experts working on solving problems is a means of collecting domain knowledge without the cumbersome exploration of texts, handbooks and other formal sources of expert knowledge that is required in constructing the first module (the so-called cognitive base) of any Expert System (ES). Through direct observation of how experts go about solving problems, domain heuristics in a sense become easily detected, which allows a quicker construction of artificial reasoners. Participant observation of expert problem solving can be also seen as a method for pre-structuring problems or as a tool to elicit cognitive chunks in a wider effort towards knowledge acquisition (KA). Characteristics and the 'technical' needs of PO were discussed in a seminal work by Forsythe and Buchanan (1989) and have been discussed in a number of works by our group that draw directly on that fundamental contribution (Barbanente et al., 1993b; Borri, 1994).

Even if recent outcomes of our research seem to call for the integration of expert individual problem-solving procedures into a more complex, hybrid, knowledge framework to overcome knowledge generalization/validation asperities and inconsistencies, participant observation meetings will probably conserve their important role in KA for the time being.

We formulated the problem at the core of our experiment as follows: on the basis of a map of a town, make a hierarchical-functional classification of roads in the light of the territorial-urban pattern; point out road restructuring needs as they emerge from the observation of relationships between land-use and characteristics of the road pattern; lastly point out potential conflicts between those needs and environment preservation demands.

We selected the two experts from the faculty of the Department of Transportation Engineering of the Polytechnic of Bari, Italy. Expert A is a 55-year-old associate professor of Road, Railway and Airport Special Constructions, a graduate in civil engineering who sometimes acts as a consultant professional. Expert B is a 52-year-old associate professor of Land Traction and Transportation Systems with a subsidiary teaching post in Transportation Systems Planning; a graduate in civil engineering like Expert A, he also acts as a consultant professional [5].

The main map shown to each expert was an aerial-photogrammetric representation on a scale of 1:5000 of a town of about 11,000 inhabitants (1991 Census) in the area near Lecce, Apulia. To allow the experts to gain a better understanding of the area surrounding this town, a 1:25000 map was also available. The zoning scheme of the master plan (a 1:5000 map) was also shown to the experts, towards the end of the meetings, to offer a better understanding of urban land-use pattern.
The PO meetings showed clear that neither expert had any personal experience of the specific town setting.

Expert A immediately identified the main arterial road shown on the map ¾ "mmmm, here the immediate reading is sufficiently easy, at least in its essential features ...." ¾ seemingly by using a combination of locational and historic understanding to fit road pattern elements and axes into the complex map scenario. His perception of other minor roads on the urban fringe and at the core of the urban built space, of railway and bypass artery axes on either side of the town, immediately followed.
Instead of going directly on with the hierarchical-functional classification of the road system, he evidently preferred to make some general considerations on transportation functions. He emphasized the need, in solving the problem and in contextual analysis, for a territorial map at least on a 1:25000 scale that covered the area outside the municipality boundaries, for he also said that road classification, in the transportation field, generally lies on both functional and administrative assumptions; he used the gravity paradigm to point out flow and density characteristics of urban system interplays.
He stressed the importance of formal direct analysis to assess traffic volumes and to support the use of intuition and qualitative and informal notations with quantity data. In his opinion, activity pattern analysis, both at local level, and at interplay level with urban centres in the surroundings, would give a sounder sense of adequacy/inadequacy dialectics in the existing local transport infrastructure supply.
At this stage, once the above-mentioned general arguments had been dealt with, he seemed to forget and/or underestimate the road classification task facing him, and gave attention to the barrier-effect of the local railway along the entire southern side of Campi Salentina (there is practically no part of Campi's urban settlement to the south of the railway corridor) and point to the need for specific railway crossings with better technical solutions than the existing ones. Also at this stage, he seemed to be aware of the lack of a direct road link between Campi S. and Veglie (a town of 13,000 inhabitants about 10 km away) and argued for the creation of a new direct link between them that would reduce traffic on national road n. 7 (the main artery he had identified when the experiment started).

He then discussed the role and characteristics of the bypass artery on the northern fringe of the core district of Campi S. (a semi-circuit which today partially crosses the urban area, due to the existence of an illegal residential district on its extreme northern fringe), and proposed a new southern semi-circular artery, to complete the existing bypass scheme and connect radial axes linking Campi S. to its southern surroundings beyond the railway.
He emphasized the importance of the biunivocal relationship between transportation (i. e. roads) and production; as for land-use and activity pattern analysis, in the town, he said

that the situation shown on the map is clear and typical of places about the same size and pointed out mainly the interplay between road patterns and residential (old and new) patterns of its urban area.
Apparently reaching the end of his problem solving - but in response to specific repeated requests by the three observers not to forget the last part of his task (potential environmental conflicts arising from new infrastructure matching transportation needs) - he argued that local morphology is favourable (there is a plain and no physical barriers); that one certainly has to know the local situation in detail as regards possible protection areas and "valid things you can find locally"; that the modest exigency of new transportation infrastructures coming from the sketchy analysis of Campi's situation would not cause real environmental risks and the use of scarps (*terre armate*) for possible elevated axes could be a satisfying way to avoid possible damage. He recalled having seen valid examples in the Trentino and Umbria regions of the need to preserve local building traditions: for example the dry walls of the Apulian countryside as land enclosures).

He seemed particularly sure that no environmental risk was linked to the possible local transportation policies. He ended with an exhortation to see the existing railway as a good opportunity to enhance mobility efficiency in the Campi area and to introduce zoning and transportation schemes into the master plan and more generally into local urban policies so allowing a better incorporating of the railway corridor into the urban area.
In contrast to the intuitive start by Expert A, Expert B showed in the early stage of problem solving routine a quasi-mystic approach, self-confident, far from saying at once that "things are clear" (a phrase that seems to be the typical start of many expert solving performances: see, for example, the geologist's verbal protocol in a classification task as reported by Barbanente et al., 1992). He spent about 15 minutes silently looking at the Campi urban map and also asked the observer to turn off the tape recorder until he had confirmed he was ready for the experiment.
In this early stage, he seemed uncertain of the scale of the map, and measured some major axes of the urban built space. He explicitly articulated his problem-solving routines in three or four items, apparently in parallel to the way the problem was formulated.
The first major issue he stressed out loud was the importance of interplay between land-use and transportation - with particular attention to the dialectics between public and private transport - but the observer could note his attention to the railway and its role as a barrier to urban development was his real starting point.
He argued that poor utilization of this major public transport means usually entails excessive reliance on cars; moreover, if the urban area "escapes" from the vicinity of the railway - as it has done at Campi - urban distances get too long, given that the core of the public transport was not located at the centre of gravity. He said his traffic-plan policy was to optimize the location and use of the railway - by re-centring urban parts around it - which is one of his recurrent themes.
The second issue he pointed out was especially private parking. Many Apulian towns, he said, having rural origins, still have buildings with large ground-floor rooms, originally used for horse-drawn carts and now used as (parking) garages, and he thought it possible that Campi Salentina also had this asset.

He seemed to pay greater attention at this stage of the problem solving routine than Expert A to the urban pattern; he noted, for instance, the inadequacy of public parking lots in the whole urban area and, in general, the lack of open spaces at the historical town core. He said that for him an important element in assessing parking needs is to look at the 'dirtiness' (for him, cars equal dirt) or 'cleanliness' of roads.
His third issue was the hierarchical classification of the road system. He envisaged a four-level classification in principle of streets (the flowing arterial to local district) but noted, however, the absence in Campi of real fast-flowing axes (excepting the modest northern bypass) and of real arterial axes (he said he could see real arterial roads only in the territory outside the urban area).
He said a closer look at the road pattern revealed a second - southern - semi-circular bypass that touched the railway in the railway station square. He said that, although part of it runs through built-up areas, this quasi-bypass could be important for the town's transportation needs, but he emphasized the need to further bypass the town north and south, in case of future development on the urban fringe taking place, which it is doing but very gradually. He said that where "natural" paths are lacking, and putative roads are "dirty" with parked cars, friction between mobility and land use should be overcome by new interventions, also by boosting public transport.
Urban distances, in Campi, do not seem to call for the expansion of public transport, even if a first line could be envisaged as running through the southern bypass. As for district roads the situation in Campi is satisfactory because of the good density of axes crossing the urban area in the two perpendicular directions .... Perhaps a north-south district road, tangential to the core, could also serve as a fluidification lane, given its relative width.
Expert B seemed to understand the urban pattern better than Expert A as an arrangement of four 'quadrants', each of them with its barycentre, crossed by two fundamental axes.
As for the lowest, local level, Expert B pointed out some disfunctions that could arise from Campi's compact historical core district but reserved judgement. He gave some attention to social trends in the use of urban space, especially in the urban core of leisure activities, and pointed out the potential need for pedestrian zones; moreover, on the fringe of the north-east quadrant we might deduce - from the bright lights of the urban core - numerous cars without parking space: for these there is a need to supply further parking areas, which, in turn, will give rise to friction, in particular when it is mainly young people who converge on the urban core (even if they tend spontaneously to shift their leisure locations) ....
Towards the end of the session, Expert B remarked on the importance of directly surveying parking needs by walking about the town at all hours (early in the morning to observe working people's mobility, late in the evening to observe young people's activities) to get a good sense of realities: "dirty" or "clean" streets, lack or abundance of private parking and/or garages to gauge unsatisfied or satisfied parking needs. The walker should also consider whether cars belong to people who own or rent their dwellings and keep in mind that as tenants generally dislike buying parking space for their cars, this tends further to complicate an already complicated situation.

10.5 Conclusions

Environmental interplay between transport and land-use has not yet been discussed adequately in published work. Even the principle of sustainability has not been described univocally: approaches to it depend on the assumed concepts and goals and/or values related to equity, environment and development issues.

Most researchers would agreed that more work is needed to see how to integrate ecology and a given economy rather than accepting the mere market as a means of indicating strategies. Such integration is not a simple linear process as it refers to the risky, irreversible, unpredictable characters of phenomena and actions.

The literature maintains that, in theory, planning can do much towards achieving sustainable development, by relating sectoral or local interventions to a broad, long-term programme that allocates action to achieve global sustainability to relevant spatial and social levels. But in fact little of the sort is yet clear. Participatory planning dialogues is surely one way to proceed, but some cases of conflict and competition suggest an institutional interplay that is not exactly a dialogue: as Section 2 of this paper suggests, transportation planning is paradigmatic.

What is more, the recent spatial-temporal dynamics of post-modern society increasingly make interaction between transportation and the spatial organization crucial, in the light of dialectics between economic and environmental needs.

As to experimental work, observations of ways to solve transportation problems would be useful in assessing whether current cognitive models, methods and strategies are adequate or not. The transportation planners who contributed to our experiments concerning a small Apulian town demonstrated some narrow technical approaches that ignored sustainability problems and conflicts between land-use and transportation. If current planning routines employ behavioural analysis and patterns referring to individual needs, preferences and aspirations, use of the simple gravity model paradigm misleadingly reduces complexity and obscures environmental and other major concerns.

The observations during our experiments suggest that experts' intuitive assumptions derive not from any deep, formal analysis but from a "sense" of the urban spatial/social context and (in our case) seemed to lack any great concern for environmental matters. The observed abundant flows of micro-evaluation and micro-decisions accord with the outcome of other recent work by the present authors on knowledge acquisition and representation (Barbanente et al., 1993a).

References

Archibugi, F., Nijkamp, P. (1989) (eds) Economy and Ecology: Towards Sustainable Development, Kluwer, Dordrecht.

Barbanente, A., Borri, D., Conte, E., Maciocco, G. (1993a) "Planning urban open space by expert system" in Beheshti, M. R. and Zreik, K. (eds) Advanced Technologies, Elsevier, Amsterdam, 155-164.

Barbanente, A., Borri, D., Pace, F. (1993b) "Microproblems and microplans in artificial reasoners" in Klostermann, R. (ed.) Proceedings of the Third Conference on Computers in Planning, Georgia Tech, Atlanta.

Barbanente, A., Borri, D., Esposito, F., Leo, P., Selicato, F. (1992) "Automatically acquiring spatial concepts by artificial intelligence planning techniques" in Campari, I., Frank, A. W. (eds) Concepts and Methods for Spatio-Temporal reasoning, Lectures Note in Computer Science, 239, Springer-Verlag, Berlin.

Bertuglia, C. S., Lombardo, S., Occelli, S., Rabino, G. (1993) "Innovazioni tecnologiche e trasformazioni territoriali: Il modello TELEMACO", in Proceedings of the XXIV Italian Conference of the Regional Science Association, Bologna, October 1993.

Blowers, A. (1993) "Environmental policy: The quest for sustainable developments", Urban Studies 30 (4/5) , 775-798.

Borri, D. (1993) "Formal versus informal procedures in planners expert system" in Khakee, A. and Eckberg, K. (eds) Process and Policy Evaluation in Structure Planning, Swedish Council for Building Research Working Papers, University of Umea, Stockholm

Borri, D. (ed.) (1994) Artificial Intelligence for Spatial Planning, Kluwer, Dordrecht (forthcoming).

Braat, L. (1991) "The predicting meaning of sustainability indicators", in Kuik, O., Verbruggen, H. (eds) In Search of Indicators of Sustainable Development, Kluwer, Dordrecht.

Breheny, M. J. (ed.) (1992) Sustainable Development and Urban Form, Pion, London.

Castells, M. (1989) The Informational City. Information Technology, Economic Restructuring and the Urban-Regional Process, Basil Blackwell, Oxford.

Fareri, P. (1990) "I sistemi innovativi di trasporto, ovvero il difficile avanzamento di decisioni sovraccariche", Urbanistica 100, 98-105.

Forsythe, D. E., Buchanan, B. G. (1989) "Knowledge acquisition for expert systems: Some pitfalls and suggestions", IEEE Transactions on Systems, Man and Cybernetics 19 (3), 435-442.

Giuliano, G. (1989) "Research policy and review 27. New direction for understanding transportation and land use", Environment and Planning A 21, 145-149.

Hart, T. (1992) "Transport, the urban pattern and regional change, 1960-2010", Urban Studies 29 (3/4), 483-504.

Harvey, D. (1989) The Condition of Postmodernity, Basil Blackwell, Oxford.

Hayer, M (1992) "The politics of environmental performance review: Choice in design", in Lykke, E. (ed.) Achieving Environmental Goals, Belhaven Press, London.

Hunt, J. D. and Simmonds, D. C. (1993) "Theory and application of an integrated land-use and transport modelling framework", Environment and Planning B 20, 221-244.

Macchi, S. (1993) "Il rapporto ambiente-sviluppo nella pianificazione del territorio", XXIV Italian Conference of Regional Science, Bologna, October.

Maciocco G. (1991) La Dimensione Ambientale della Pianificazione del Paesaggio, Angeli, Milano.

Maciocco G. (1994) "Sistemi intelligenti e pianificazione urbana: Alcune riflessioni", in Maciocco G. (ed.) La Città la Mente il Piano, Angeli, Milano.

Mackett, R. L. (1990) "Comparative analysis of modelling land-use transport interaction at micro and macro levels", Environment and Planning A 18, 307-322; 511-528, 22, 459-475.

Merlin, P. (1984) La Planification des Trasports Urbains. Enjeux et Méthodes, Masson, Paris.

Mitchell, R. B., Rapkin, C. (1954) Urban Traffic: A Function of Land Use, Columbia University Press, New York.

Newman, P. W. G., Kenworthy, J. R. (1989) "Gasoline consumption and city", APA Journal 55 (1), 24-35.

Owens, S. (1992) "Energy, environmental sustainability and land use planning" in Breheny, M. J. (ed.) cit.

Plotkin, J. (1987) "Property, policy and politics: Towards a theory of urban land use conflict", International Journal of Urban and Regional Research 11, 382-402.

Plowden, S. (1985) Transport Reform: Changing the Rules, Policy Studies Institute London.

Report on Traffic in Towns (The Buchanan Report) (1963), HMSO, London.

Rho, J. H. and Kim, T. J. (1989) "Solving a three-dimensional urban activity model of land use intensity and transport congestion", Journal of Regional Science 29, 595-613.

Saunders, P. (1981) Social Theory and the Urban Question, Hutchinson, London.

Scandurra, E., Macchi, S., Goffredo, G. (1993) "Innovazione telematica e nuove centralità: Il caso del lazio" in Lombardo, S., Preto, G. (eds) Innovazioni e Trasformazioni della Città, Angeli, Milano.

Secchi, B. (1973) "Sviluppo della rete di trasporti e distribuzione territoriale degli insediamenti in italia", Urbanistica 61, 88-93.

Tolley, R. (ed.) (1990) The Greening of Urban Transport, Belhaven Press, London.

V. A.(1973) Urbanistica 61.

V. A.(1977) "PRG e piano dei trasporti e della viabilità di Modena", Urbanistica 66, 5-38.

Van der Valk, A. (1991) "Growth regions. A new perspective for urban and regional planning in the Netherlands", Annual Conference of the British Section of the Regional Science Association, Oxford, September.

Vittadini, M. R. (1991) "Il fascino discreto delle metropolitane leggere", Archivio di Studi Urbani e Regionali 41.

Vittadini, M. R. (1992) "Infrastrutture contro la città" in Salzano, E. (ed.) La Città Sostenibile, Edizioni delle Autonomie, Roma.

Wegener, M., Spiekermann, K. (1993) "Comprehensive land use/transport models and the ecological reorganization of metropolitan regions", Proceedings Third International Conference on Computers in Urban Planning and Urban Management, Georgia, Atlanta.

Witulsky, K. (1992) "Knowledge based route selection in public transport", Second International OECD Workshop on Knowledge-based Expert System in Transportation, Montreal, Quebec, Canada (unpublished).

Zambrini, G. (1985) "Un'area di incertezza", Urbanistica 78, 6-16.

Notes

[1] This paper is the result of the authors' team work: their contributions are as follows: Section 10.2 by V. Monno, Section 10.3 by A. Barbanente and Section 10.4 by D. Borri.

[2] In saying this, we obviously do not intend to imply that scholars gave little attention to the two-way interaction between mobility and spatial organization: it is enough to think of the great success that spatial-interaction models (derived from the Lowrian model aimed at defining optimal activities and land-use patterns) have had with regional scientists (to refer only to some recent contributions, see, in the wealth of literature, Hunt and Simmonds, 1993; Rho and Kim, 1989; Mackett, 1990; Prastacos, 1986); and we do not lack examples in planning, also at urban level, in which mobility and spatial organization have been assumed as central elements of land-use choices (from the well-known master plans for Copenhagen and Stockholm, and, in Italy, from the experience of the Municipality of Modena, perhaps the first example in Italy of mutual interaction between general town planning and transportation planning).

[3] We should consider that, since the late 1950s, the motorway works sector has been mainly a concession to the state-controlled industrial company IRI, and that, earlier, before the state itself promoted a support policy for the car industry, the role of big private companies was direct (the "Autostrada del Sole Milano-Napoli, between 1952 and 1954, was designed by SISI, a company that included AGIP, FIAT, Pirelli, Italcementi; on these events see the monographic volume on transportation of *Urbanistica*, 1973).

[4] From our point of view, the problem when the 'external' intervention has to be evaluated is not the lack of conformity with the plan, as a plan in the present Italian situation is a 'tool' that rarely constitutes a frame of reference representing actual development goals of a certain local community. The real problem is a lack of a range of criteria and rules as a basis on which to evaluate spatial modifications caused by the proposed transformation in the light of the interests, preferences and aspirations of the communities that will suffer most directly from it.

[5] P.O. meetings were held at the Department of Town Planning with A, and at B's home. The first was attended by three observers, the second by one observer. Both the meetings took two hours, including the time needed for introduction, diversion pauses and the like.

Angela Barbanente
CNR IRIS
Strada Crocifisso 2b
70126 Bari
Italy

Dino Borri
Valeria Monno
Dipartimento di Architettura e Urbanistica, Politecnico di Bari
Via Orabona 4
70125 Bari
Italy

11 CRITERIA FOR CHOICE AND EVALUATION PROCEDURES: THE CASE OF URBAN TRANSPORT INFRASTRUCTURES

Donato Caiulo
Francesca Pace
Francesco Selicato

11.1 Introduction

This study stems from the technical-political debate around the building of two important urban roads that were intended to have considerable effects on the development of the two towns involved, Bari and Brindisi[1]. The argument which emerges is that, as the current economic crisis heralds an uncertain future, decisions in the transport sector, particularly about building new infrastructures traditionally intended for private-vehicle use, should be considered as essential premises for solving the problems that accompany social-economic and territorial development.
When analysing the two projects an attempt has been made to investigate thoroughly the motives behind the choices made in particular technical-administrative units such as those of medium-sized southern Italian towns with low technological standards, limited information and knowledge and reduced economic resources.
In similar situations, decision procedures are influenced because few protagonists refer to consolidated procedures based on a traditional concept of urban planning. Any occasional external evaluation of systems and models is reduced to a mere formality.

11.2 North-south arterial road in Bari

Background history

The building of the north-south urban trunk road, now almost finished but still encountering difficulties, is significant in a number of ways.
The road is one of the principal ones included in the General Town Plan for Bari and cuts through the town from the port to the hinterland. The first section of this fast trunk road with a hard shoulder includes building of a big circular branch road and a road branching off at right angles (in an east-west direction). It is financed in accordance with law no.64/1986 (*Intervento Straordinario per il Mezzogiorno*).
Its significance is determined by conformity with local authorities' policy making and is generally typical of towns in South Italy: "off the budget" economic resources; using

D. Borri et al. (eds.), Evaluating Theory-Practice and Urban-Rural Interplay in Planning, 141–154.

funds allocated for particular occasions (in this case the World Football Championships in 1990) and for other ends than those originally stipulated; the influential presence of local building entrepreneurs in decision taking; and hurried discussions influenced by inescapable factors (availability of funds and therefore creating new jobs). In this case the large amounts of money brought in by powerful local politicians and entrepreneurs to further urban transformation projects which, even when planned (the North-South road built in conformity to the General Town Plan had been approved as far back as 1976) would require considerable time to be set up.

The starting point was that Bari was to one of twelve Italian towns that would host the championships. The local authorities decided on a group of mainly public infrastructure works to be among the urgent interventions to be financed; there was little interest in them, for virtually all local interest was in building a new stadium (which was considered a determinant factor for hosting of the manifestation). Operations for the stadium (begun July 1987) met with hostility from the political opposition and the Greens initially because of the large financial resources required when the town presented other pressing needs, and later because historical-archaeological remains came to light in the areas to be used.

The siting of this stadium was decisive for the decision making procedure and for establishing a fixed scenario. The area chosen was one of three alternatives indicated in the General Town Plan. It take places to the south of the town but it had poor accessibility (there exists only one narrow road, SS.271 Bari-Matera, that has a heavy traffic load). It lies near a branch road enclosing a circular area of services and administrative offices and has been indicated by the local authorities as a possible site for the Regional Council's head offices (the political opposition has proposed another site to the west) that would favour urban building developments south of the town where considerable land investments are concentrated.

The north-south road proposed as necessary for the new stadium is the only completely newly built road decided by local authorities bodies among other works, mainly road infrastructures intended to improve access to the town from the outskirts and also mainly for private vehicle use. The funds allocated amount to 198 milliard Liras to be added to the sum of 55 milliards allocated in 1987 for building the new stadium.

The transfer to the south of a number of services and administrative offices made a new highway indispensable; building the new stadium provided funds for this and also speeded up the operation.

The north-south road costs 40 milliard liras plus 25 milliard for the east-west branch. It does not reach the port but stops a few metres from the railway line under which a subway is being built to be connected up with another road; the project has reduced the width of the north-south road to 25 metres and does away with the separation between land and shoulder running at ground level and with a number of intersections. Thus the infrastructure no longer resembles a motorway connecting the hinterland and the port also for the use of heavy vehicles but rather interacts with an already developed consolidated urban context.

The road runs through built-up areas and open land and crosses railway tracks and is superimposed on the existing town without taking into account the environmental features of much of the undeveloped land: the gullies and rock caves for the protection of which the Superintendency for the National Environmental, Architectural and Artistic

Heritage has recommended that "the road system should be compatible and to this end in conformity with the Act no.1089 of 1939, have declared the area a protected area and suggest it become an archaeological park."

Figure 11.1: Map of Bari

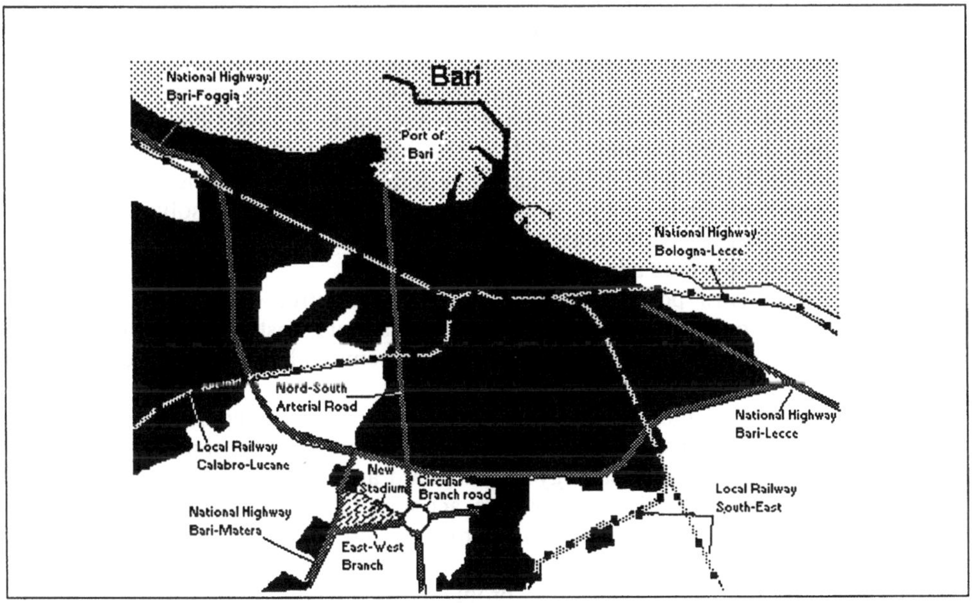

In building the North-South trunk road concurrently with the new stadium, the first section of the road is still incomplete. The plans for the road were drawn up in different phases and did not follow a regular pattern. Initially the work was opposed by the left-wing parties, in particular the Greens, although they disagreed among themselves, and the parties anyway intervened after a decision to start had been taken. No public discussion took place of the reasons why local administrators favoured the north-south road, which were that this would make better use of the stadium and conform with the General Town Plan.

Only after the decision had been taken did previously merely marginally involved groups emerge to participate in the discussions; for example, ecologists raised environmental problems and requested an impact assessment report only after tenders had been submitted. Local Administration authorities asked constructors submitting tenders to include an impact study when the altimetric and planimetric aspects of the layout had already been stipulated in the working project[2].

The matter is far from being concluded: in the absence of mediators, administrations, superintendencies and environmental groups are clashing without intending to find possible alternative solutions and are thus holding up the building phase. All this comes on top off when procedural reasons are blocking the financial contribution for the second section of the road from the railway to the port, so that the section now undergoing construction is of very little use.

Protagonists and strategies

Conformity to the General Town Plan that the administration authorities put forward as a formal instrument for legitimizing intervention (Mazza 1992) is a factor that carries considerable weight with public opinion. The plan is seen as firmly established with its normative, regulatory aspects as main reference points.

Here we should point out that most of the objectives underpinning the Plan for Bari dating from the end of the 1960s (the first draft was made in July 1968) have not been met: they include its dimensioning, which is absolutely out of proportion, its role as a town-region, its emphasis on the infrastructure system, and its town-design approach. But this has not been considered enough to warrant[3] a revision aimed at solving the town's new emergencies: to recuperate most of its central areas, to create green areas and to distribute social services more adequately.

Implementing the Plan's policies seems beyond the control of the community and consequently of the political and social groups that should, institutionally speaking, be responsible for implementation. Attention generally seems to be centred on temporary or even marginal matters, not on its far-reaching aspects.

The position of some of those involved in making decisions, although consistent with their roles and declared objectives, is not always consistent with the timing and modality of action. No long-term general strategy to safeguard and use the area's rich natural and historical heritage was established, nor was account taken of residents' and owners' claims that emerged only when work was in progress and the contractors were providing for expropriation.

A deeper study of substantial issues (the problems of intervening in a densely built-up area; long, costly expropriation procedures; the presence of historical remains throughout the area; the need for a completed assessment of the town's everyday requirements and objectives) would certainly have led to alternative solutions.

11.3 Brindisi: the Pittachi Road

The planning and initial building phases of the *Pittachi* 'districts' road have to be included in the urban-administrative facts of the town of Brindisi before the implementation of the General Plan (adopted in 1980 and officially implemented in 1989). In 1988 the local authorities suggested some seventeen projects and feasibility studies to the regional authorities to be financed as stipulated by the Act 64 of 1986: only the Pittachi road project was passed.

This important urban artery was financed in March 1990 to the amount of 25 milliard Liras (the original project stipulated 37 milliards) with the requirement that the plan should be re-worked in the light of the reduction in funds. In two separate, unanimously approved resolutions in June 1990 and September 1991, the commune assigned the work of re-drawing the plans and executing them to the contractors.

Figure 11.2: Map of Brindisi

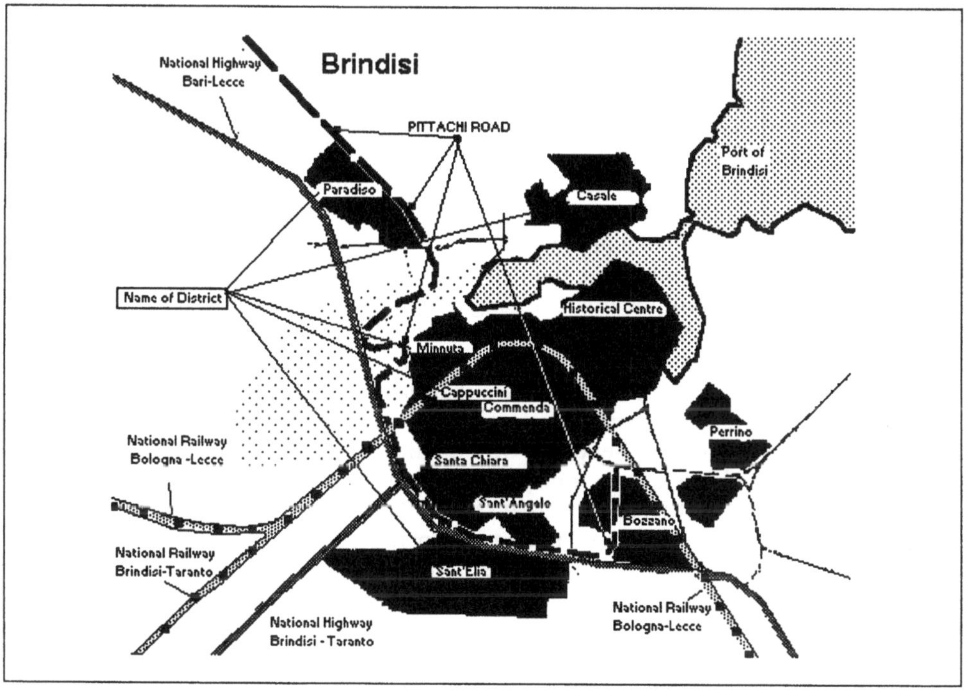

The re-worked project, and a relative cost-benefit estimate, was approved under the strategy of "the requalification of urban and historical centres". In addition to the Pittachi road, eight of the projects proposed for funding according to the 64/1986 law involved urban and building requalification; five of them concerned infrastructure road works excluding the *Pittachi* Road, in any case geared to the requalifying of the urban context, lastly two proposed studies on the definition of informative systems[4].

The re-worked road project comprises a carriage way about 2100 m long that, with some minor works (branch roads and re-paving work) amounts to about 3200 m of carriage way. The 10-m wide carriage way comprises two 3.5-m lanes with a 50-cm wide bank along on one side and a 2.5-m hard shoulder on the other side. The road begins north of Brindisi, in the Paradiso quarter, follows the west bank of the Cillarese canal, crosses it near the SS 379 highway and then follows the east bank to connect up with Via del Lavoro.

The plans for a bicycle lane grew so important in the project almost to the point of justifying the plans for the main road itself. The lane was to be 3m wide and 5500m long and to run beside the road but diverge from it at certain points and finally run between the two planned *Cillarese* and *Patri* parks.

With the building of this road, the General Town Plan reversed a trend exemplified by the settlements that had grown around roads radiating away from the town centre. This new road and the settlements set up along it was conceived and promoted to define linear town development[5], with a series of districts set up along it in a north-south direction. In the plan, parks and green belts would separate these districts that were to be supplied with all facilities and positioned so as to avoid their degeneration into conurbations, as has happened along the main roads into the town (*Via Appia, Via S. Vito,* the road to Lecce).

The plans for this road also intended to avoid intensifying streams of traffic going from the centre to the outskirts: it would directly link the districts on the outskirts and connect them to the port and the industrial area but not via the centre. Any future settlements or areas for development were to be situated along this road which would therefore become of fundamental importance in the urban contest.

The General Town Plan emphasizes the different objectives to be reached by including such a road on the borders of the town: among them are control over dynamic expansion; the morphological lay-out; and the control of living quality (with reference to environmental and cultural standards). But evaluations here have not explored how an infrastructure of this kind can influence the urban layout, its transformation, persistence or modification and hypotheses as to future urban developments.

Besides being included in the General Town Plan, this road is also foreseen in plans along economic - social and urban - transport lines at levels higher than the local level. Its construction and main function are discussed in the "Feasibility study of the Italian-Greek trans-frontier project for the completion and expansion of the route between Europe and the Middle East"[6] and in the "Special Project for the Ionian-Salentine urban system"[7]. Especially in the first, its building represents a complementary phase in the setting up of an international route Brindisi-Igoumenitsa-Volos. Both are ambitious projects and the specific urban objectives behind the building of the Brindisi road do not seem to meet their requirements.

11.4 From decision taking to realization: comparison of the two cases

In both cases the planning solutions stem directly from strategic decisions taken within the General Town Plan.

The Pittachi road is planned to re-design the outlying border districts of Brindisi; some are already built up, others remain to be settled, and this road is intended to absorb traffic that currently crosses the town centre. The specific objectives of building this new artery are to relieve traffic pressure on the roads out of the centre (in areas where settlements have been concentrated over the last ten years); to ease communications between the outlying districts; and to `merge or coagulate' (as the report says) the new districts set up along the road; the more general strategy, which is harder to justify but frequently emphasized, is to allow Brindisi to play an important role in the territory[8].

The north-south trunk road in Bari for fast traffic from and to the town's southern hinterland conforms well formally and functionally with the plan's leading idea, now seen as unrealistic, of conferring the status of city-region on this regional centre.

In neither case, then, are legitimization and consensus to be found at the level of macro-territorial objectives, and particularly not for Brindisi where the specific objectives themselves are sufficient to warrant consensus and where the general objectives and the plan do not seem to match. Drawing attention to general rather than specific factors is a recurring practice when seeking consensus for projects. It is a fact that when a plan mentions policies of a symbolic character (the city-region, the Ionian-Salentine city) this often ends by masking the real policies by conveying an impression of a "comprehensiveness which dilates the features of the plan so as to reduce the conflict around the real motivation behind the decisions taken: if nothing else this reduces opposition to the plan" (Mazza 1986).

A close look at the planned solutions suggest that in Bari the prevalently built-up territory that has numerous environmental assets, is considered a mere container for the proposed - but totally foreign -interventions, while in Brindisi where urbanization is less diffused but has an equally significant morphologic configuration, the plan describes the road as intended to recreate the "global image of the landscape as seen from the west", which is what prints show[9]. Environmental issues are totally ignored in Bari but cautiously advanced in Brindisi, but only pictorially. The towns' plans are far from conferring the environmental system in either with an ability to "sustain" man's intervention without having its order and components upset (Howe 1979); instead, they focus their interest on the changes to be made. Thus the formal elements of urban planning predominate in both solutions: the straight line of the north-south road, the curving line of the Pittachi road following the Cillarese stream.

Belonging to an outdated generation (Campos Venuti 1987), the two plans are penalized by a grave technical drawback (Balducci 1991): they lack a systematic knowledge of the whole natural and human environment and of the deep-rooted structures of environmental processes through which may be discovered "the quality of the differences of its structural features" (Maciocco 1991). Such knowledge would promote planning action by offering a means of dealing with the environment seen as background. Cognitive action would thus become "planning interpretation" (Palermo 1993).

Conversely, in Bari (and perhaps also in Brindisi) knowledge is being built up step by step as work progresses, mainly because indications that are followed by investigations and checking procedures lead to protection action being begun. These procedures seems to impart a warning rather than issuing order for all they result in is that the Superintendency recommends that the roads be built in a way that safeguards the environment[10].

It should be said that, in Bari, and only after external pressure, a partial recovery of environmental values was provided by the EIA procedure when the tender had been decided. It has already been observed that one of the most important phases in the planning process for infrastructures is the preparation of the plan-project, meaning a plan that, alongside the presentation of infrastructure alternatives, indicates the project aims, the interests to be protected and the problems to be solved (Linden 1989, Voogd 1993); but just as evidently, as in the present case, and granting the need for specific evaluation procedures, while the "traditional" plan still legitimizes choices, evaluations and changes, a set of reference parameters (even an executive plan, such as the one for north-south road that was the basis for tenders) can be defined but without allowing for any variation, so that judging the work as suitable automatically means judging its conformity to the plan (Mazza 1993).

Both the roads favour motor, and essentially private, vehicles, despite high levels of air pollution impose restrictions on circulation almost everywhere; Bari was one of the first communes to take action in this sense. The *Pittachi* road, although it diverts vehicles from the centre, can only reduce concentrations of carbon monoxide. The benefits, to be considered as such, must be seen in relation to the number and variety of involved groups and to the greater awareness among future generations of ecological problems, which promotes inter-generational equity (Fusco Girard 1993, Howe 1993).

In connection with project solutions it should further be pointed out how the plan's technical limits also constitute limits to the project during its implementation because the goals that justified the choice of those project solutions are no longer valid: thus the north-south road is drastically narrowed by eliminating the separation between fast lane and hard shoulder while, contrary to the original plan, it runs at ground level with a number of intersections. The *Pittachi* road, on the other hand, apparently more on account of cost than of physical factors is, in the end, more like a normal unimproved road with a hard shoulder, while its bicycle lane, even though an urban novelty in southern Italy, does not seem particularly significant.

In both cases the declassing of functions is accompanied by the absence of the development and/or building of a smaller network able to ensure that the urban components are well connected (Giordano 1990), because much time will be needed to complete the roads, although especially the Bari road will probably remain permanently incomplete. As regards the question of connectivity it should be pointed out that for neither road the possibilities of re-qualifying the existing road network have not been explored so as to make it more accessible.

The north-south road was built not to meet general territorial demands but essentially to take advantage, including financial advantage, of the building of the World Championships stadium. Although planned within a general urban plan the executive phase and its relative goals were occasional. In Brindisi, the plan set the priorities for the interventions and their financing. With the objectives of the *Pittachi* road being mainly fixed in a macro territorial dimension, the plan gave it priority as an infrastructure intervention while stipulating that its building costs should be met in part from urban taxation because it had a mainly urban function[11].

The road project, together with other projects, was approved by the commune some months before regional approval was obtained. However resources were made so quickly available that it could be financed entirely with state funds in accordance with the Act 64/86. The project is part of wider organic action to "requalify urban systems and revitalize internal areas" so underlining the urban valency of the work. The building phases of the north-south road are appropriately inserted in a different strategy, within the ambit of "interventions on transport systems" also financed by the Act 64/86. It is interesting that, unlike what occurred in Brindisi, the public administration was totally absent from the planning phase, while a group of businessmen, motivated by their own interests, showed marked entrepreneurial capacity.

It should be emphasized that expropriation procedures in both towns were left to private interests: in Brindisi to the company responsible for the executive phase and in Bari to the contractors. This contributed to social tension in both towns, by exposing the readiness of public authorities to delegate their duty to safeguard all citizens' rights. This comes from the very evident lack of entrepreneurial culture among southern public authorities: they see entrepreneurial undertaking and decision making as dimensions of responsibility rather than of roles (Senn 1986, 1988).

Uncoordinated action in environment and transport matters often leads to conflicts, as the foregoing clearly shows, and is often accompanied by social tension as a consequence of poor results and high social, economic and environmental costs.

11.5 Evaluation procedures

Among the many articulated definitions of Environmental Impact Assessment (EIA), references to the "interpretation and communication of information pertaining to eventual impact" (Munn 1979) are very frequent and its utility is seen essentially "in communicating to the community and to decision makers the advantages and disadvantages of the proposed work identified on the basis of pertinent criteria, in comprehensible terms" (Unep 1978).

Actually the communication and interpretation of the knowledge phase is totally absent, no evaluation criteria for impact would appear to have been settled nor, when the experts have drawn up the different studies, have their criteria been made known. The inclusion of these, when the decisional process was already under way or had come to an end, made zero any option to unfeasibility, as all studies were directed to justify the technological and building choices made, and the measures needed to describe how to deal with the effects of expected impacts.

At no time during the decisional process was the feasibility of the proposed planning model verified: the interaction of the road artery with the general infrastructure system, the possibility of putting forward alternatives such as the strengthening of other forms of transport or road arteries.

The lack of public debate and a true decisional arena prevented (or in some cases evaded) the holding of genuine discussions of opposing points of view, probably due partly to a distorted use of the EIA procedure and partly to its inadequacy as an essential element in discussion and/or decisions about urban transformation intervention on a wide scale.

The serious limits of Italian ruling on the EIA procedure which has taken up the EEC directive set up in 1985, but only formally, constitutes one but not the only hindrance.

Once discussion of this directive started in 1985, questions have been raised as to the competence and "suitability" of public administrations in handling the problems and consequent interventions connected with environmental issues.

With state, regional, provincial, communal and inter-ministerial sectors all involved, interventions in territorial planning often overlap or are discontinuous, and so combining competence would perhaps have been advisable[12]. Besides rationalizing and simplifying some working procedures, this would have been a first step in beginning a systematic approach to the environment and territory underpinning the directive and preceding working experiences.

The EEC directive follows practice in America in making the relations between the public and administration a central factor, which entails users take a direct part in decision making. Italian practice is different: in the debate on Act 349/86 (it established the Ministry of the Environment and provided norms regulating environmental issues), and after decrees had been promulgated in August and December 1988, clear positions arose based on the claim that "Italian administration structures are in no way able to admit procedures having participation as their main feature" (Greco,1982).

An EEC directive introduced environmental compatibility certification and a participation procedure along the lines of that stipulated in Italian urban-planning legislation. The Italian procedure[13] had failed over public participation[14] due to: the absence of precise information as to the hypotheses and transformation processes, poor knowledge of the territory, absence of phisical meeting point for the collection and the exchange of knowledge. Besides, the choice of programmatic, project-related, environmental reference frames of impact studies leads them into a purely normative dimension for checking the constraints as to methods and criteria for determining the environmental factors at risk.

The Italian interpretation of the EEC directive excludes any reference to project alternatives (including option zero), to induced direct and indirect effects, to medium and long-term effects, and moreover the categories of the works mentioned appear improbable or undefined.

This applies to the definition "fast communication roads", for which no Community or Italian classification of this kind exists. It is a debatable concept that seems to ignore roads in urban areas or those that join or cross areas particularly at risk (coastal lands, beauty spots, historical and archaeological sites, etc.). This lack of precision arises because the criteria for choosing the works relate completely to their definition, not to their interaction with their locality.

Given this normative frame of reference, it is no wonder that the usual procedure for evaluating impact is *ex post facto* to justify choices already made rather than to contribute knowledge to, and so aid, decision making within the procedures for implementing urban and territorial policies.

Perhaps the cost-benefit analysis requested by the Brindisi communal administration for building the *Pittachi* road relates more to the expected results but not necessarily to the morphological and environmental features that are also present here. This analysis, too, finds its justification in the recently passed (1989) General Town Plan that achieved a general consensus.

Environmental and distribution issues in connection with building the Brindisi road infrastructures are supposedly resolved within the Plan's decisional phase; the only legitimizing phase for the choices then to be implemented (the cost-benefit analysis does not consider alternatives but only the economic consequences of the work). But no environmental compatibility frame seems to have been included in the General Plan of Brindisi against which the planned transformations[15] could be measured, and no public debate was held on the most important decisions.

11.6 Conclusions

As to participation, it is clearly only marginal to the Plan both because of the methods that should be applied to achieve it, and because the issues in question are complex, being often presented along theoretical lines and in technical terms. Conflicts and greater participation are present in the single, more easily understood cases involving single individuals' interests (Healey, 1985). On the other hand, dividing the decisional process into segments can make participation more efficient: if this can more closely define points at issue or the parts of a general problem, it can reduce the number and size of the

consensus areas that must be found (Scharpf ,1981). Any such difference of opinion has a better chance of being resolved than when bigger, more complex issues underlie conflicts.
However, in current urban practice, even admitting the need to get beyond the Plan as a general tool for territorial organization, it is hard to find a similar model to justify procedures elsewhere (Mazza, 1986). The evaluation procedures often mentioned are discretional and not very convincing (e.g. the EIA procedure) or they re-introduce the classical rational model (e.g. the cost-benefit analysis) that is inadequate in addressing only partial aspects of problems.
Using evaluation procedures that, in a formalized synoptic form, can illustrate and describe the positions and motives of the various decision makers when decision-making starts can contribute much to accumulating knowledge, taking decisions, and exercising control by a community. We would emphasize that dividing decision-taking procedures into segments is welcome, because it makes for a better understanding of problems and for the participation of the public. When this is done, however, the complexity of the overall issues and the unity of the territory must be preserved; this can be done if reference is continuously made to the context, by applying multidimensional evaluation procedures that are diversified in relation to the object: those of intervention to times and methods; those of implementation to the relevant environmental and economic-productive categories. In other words, these procedures are set up pragmatically in each case and refer to an unchanging system of values that, not being influenced by any theory, relates to its contexts and serves such objective as emerge. To be successful, procedures of this kind can be set up only in a regime of widespread information and mature participation: in a decentralized system, it makes sense to let groups that are not necessarily institutionalized represent and protect the interests of various sections of society[16].

References

Amendola, G. et alii (1985) Segni ed Evidenze. Atlante sociale di Bari, Dedalo, Bari.

Balducci, A. (1991) Disegnare il Futuro, Il Mulino, Bologna.

Barbanente, A. and Pace, F. (1992) "Bari, scenari incerti, capacità realizzative, contraddizioni persistenti", in V. A. , La Costruzione della Città Europea, CRESME Roma, pp.409-470.

Botta, F. (1987) Sospetti e indizi. Per un riesame del caso Puglia, Dal Sud, Bari.

Campos Venuti, G. (1987) La Terza Generazione dell'Urbanistica, Franco Angeli, Milano.

Carrieri, M. (1989) "Le innovazioni imperfette del ceto politico pugliese", in L. Masella, L. e B. Salvemini, (eds), Le Regioni d'Italia. La Puglia, Einaudi, Torino, pp.981-1001.

Crosta, P.L. (1988) "L'innovazione nelle pratiche territoriali: non inerente, ma contingente al piano", in C. Donolo, F. Fichera, Le Vie dell' Innovazione, Milano

Dente, B. (1990) "Metropoli per progetti?", in B. Dente et alii, Metropoli per Progetti, Il Mulino, Bologna.

Etzioni, A. (1967) "Mixed scanning: a third approach to decision making", Publi Administratio Review 27.

Fusco Girard, L. (1993) "Le nuove frontiere nel campo della valutazione tra conservazione della qualità e sviluppo: estimo, economia ambientale e sviluppo sostenibile", in L. Fusco Girard (eds.), Estimo ed Economia Ambientale: le Nuove Frontiere nel Campo della Valutazione, Franco Angeli, Milano, pp.13-40.

Giordano, R. (1990) Il Ruolo dei Trasporti per uno Sviluppo Territoriale Diffuso, Milano.

Greco, N. (1982), "La valutazione di impatto ambientale in Italia: problematiche nascenti e necessità di una comparazione", in S.It.E., Il Bilancio di Impatto Ambientale, Elementi Costitutivi e Realtà Italiana, Parma.

Healey, P. et al. (1985) The Implementation of Planning Policies and The Role of Development Plans, Pergamon, Oxford.

Howe, C.W. (1979) Natural Resource Economics: Issues, Analysis and Policy, Wiley, New York.

Howe, C.W. (1993) "Le frontiere nella valutazione di risorse prive di mercato: problemi e prospettive", in L. Fusco Girard, (eds.), Estimo ed Economia Ambientale: le Nuove Frontiere nel Campo della Valutazione, Franco Angeli, Milano, pp.57-83.

Linden, G. (1989) Highway Location, towards a Framework for Planning Support, Act of International Seminar Groningen.

Maciocco, G. (eds.), (1991), Le dimensioni ambientali della pianificazione urbana, Franco Angeli, Milano.

Mazza, L. (1986) "Giustificazione ed autonomia degli elementi del piano", Urbanistica 82, pp.56-63.

Mazza, L. (1993) "Descrizione e previsione", in S. Lombardo, and G. Preto, (eds.), Innovazione e Trasformazioni della Città, Franco Angeli, Milano, pp.181-196.

Morgese, W. (1988) "Tipologie imitative di legislazione locale: leggi finanziarie e spese per progetti", Economia pubblica 18/6, pp.285-294.

Palermo, P.C. (1993) "Sistemi intelligenti per la pianificazione: una concezione non-cognitivista", paper presented in the international Seminar: Sistemi intelligenti e pianificazione urbana, Cagliari.

Scharpf, F.W. (1981) "Formulazione delle politiche ed implementazione. Problemi ed approcci", in C. Donolo, F. Fichera, Il Governo Debole, De Donato, Bari

Senn, L. (1986) "Una nuova cultura per una gestione imprenditoriale dell'ente locale", Confronti 1, pp.25-36.

Senn, L. (1988) "Politica urbana, imprenditorialità e soggetti decisionali", in M. C. Gibelli, M.C. and A. Magnaghi (ed.), La Pianificazione Urbanistica come Strumento di Politica Economica, Franco Angeli, Milano, pp.99-104.

Voogd, H. (1993) "Conservazione programmata delle risorse ambientali: utilizzazione pratica dei metodi di valutazione dei progetti", in L. Fusco Girard, (eds.), Estimo ed Economia Ambientale: le Nuove Frontiere nel Campo della Valutazione, Franco Angeli, Milano, pp.362-372.

Notes

[1] This Paper is the result of a collaboration and individuals' contributions are as follows: section 11.1 is by F. Pace and F. Selicato, section 11.2, 11.5 and 11.6 are by F. Pace, section 11.3 is by D. Cailulo and F. Selicato, section 11.4 is by F. Selicato

[2] It should be noted that the companies that included changes to the layout, including the winning company selected by the technical examining board, were excluded from the bidding by the regional administrative tribunal, after long legal proceedings, on the ground that "no changes were allowed for in the competition rules". The administrators principally responsible for this action have twice overruled the technical board.

[3] Recently the Superintendency ruled that the Saverio Lioce property was to be protected, a ruling later modified to allow the road to pass at a tangent to the farmhouse on the land.

[4] for further discussion on this matter see Barbanente, A. Pace, F. (1992) "Bari, scenari incerti, capacità realizzative, contraddizioni persistenti", in Aa.Vv., La costruzione della città europea, Cresme credito fondiario, Roma, pp.409-470

[5] Seven of these studies and projects were commissioned by public administration from external professionals, five from the commune's technical office; five (including Pittachi Road) were executive projects drawn up by companies or companies in consortium.

[6] EEC, CASMEZ: A feasibility study of the Italian-Greek transport system for the completion and realization of the community transport links between Europe and the Middle East

[7] ISMEZ: special Project for the Ionian-Salentine system, socio-economic study and study movement

[8] "within the ambit of works for the urban organisation the Ionian Salentine area, the chance to eliminate traffic congestion in Brindisi seems realistic ..." (pag 48 of the General Town Plan); "The new Pittachi main road has an important role and is strategically relevant in the linear development of the town and conforms to the objectibves of the expanding town" (pag 3 of the General Report attached to the project)

[9] p. 57 of the General Town Plan Report.

[10] pp. 51-52 of the General Town Plan Report.

[11] The road will to the merging of the new districts set up along this thus guaranteeing resources for its realization at least for the urban tracts (pag 52 of the General Town Plan Report).

[12] A single Agency for the Environment and Territory is referred to in a Bill presented to the two parliamentary houses during the debate on the reform of the Ministry of Public Works President' s Act no 616/1977

[13] The authorization released by the Ministry of the Environment applies to ten categories of works; those indicated by the EEC directive have absolute priority over others indicated by the single member states.

[14] The impact study is deposited in the Regional Office for a period of 90 days during which members of the public may consult it and submit their observations.

[15] Research into `environmental features' to be referred to when organizing space (Maciocco 1991) is relevant here.
[16] It must be admitted that, in any situation of this kind, better organized groups have an advantage (Etzioni 1967).

Donato Caiulo
Francesca Pace
Francesco Selicato
Dipartimento di Architettura e Urbanistica, Politecnico di Bari
Via Orabona 4
70125 Bari
Italy

PART III

ENVIRONMENTAL POLICIES AND URBAN RURAL INTERPLAY

12 ENVIRONMENTAL CONSIDERATIONS IN MINERALS PLANNING; THEORY VERSUS PRACTICE

Simin Davoudi

12.1 Introduction

In the last few years there has been a growing concern about the environment with a focus on the notion of `sustainable development' (WCED,1987; Pearce et al.,1989).
In spite of confusion over the exact meaning of sustainable development and disagreement over the practical ways of implementing its principles, sustainable development provides a radical, albeit flexible, concept that planners, among others, are desperately trying to absorb.
Its implications for the planing system are wide ranging: it adds to the range of issues included in the mediation of interests with which the system is engaged; it adds new criteria, new ways of thinking about relationships and consequences, and new ideals about the `good city' and the `good landscape'; it also raises significant institutional questions about `who does what', where responsibility lies for addressing environmental issues, at what level, with what agencies, and with what linkages to other policy areas.
Given the political, technical, and moral issues involved, incorporating the concept of sustainable development within the planning system is not a simple task. Yet, as Healey and Shaw (1993) argue, the planning system, and specifically development plans, are a key arena within which gathers one of the main challenges for contemporary environmental agenda, which is integration of economic, social, and physical dimensions of human existence.
Although in principle development plans are intended to provide a strategic and long-term approach to the management of environmental change, in reality their significance in the planning system has fluctuated over time. Overall, the planners' task has been to predict the demand for development and use of land and then to provide for it subject to certain criteria, including environmental considerations. Therefore, some of the dimensions of the new environmental policy agenda are embedded in the British planning system. Yet, there are others which are new to it.

12.2 Environmentalism and planning, a historical perspective

In Britain, environmentalism, as we would now recognize it, emerged in the late

D. Borri et al. (eds.), Evaluating Theory-Practice and Urban-Rural Interplay in Planning, 157–165.

nineteenth century. Over the past 130 years layer upon layer of membership organisations have emerged in response to new problems and new popular concerns. Despite the wide varieties covered by contemporary environmentalism. Newby (1990) identifies four distinct phases in their development.

The first phase was from the mid-1880s to the turn of the century. The dominant theme of this era was preservation of both natural and man-made artifacts. However, "these were to be preserved for `the nation' but from `the public' (op cit)". The second phase during the inter-war years experienced a widening of the social base of environmental concern to include the suburban middle-classes, particularly in the South-East of England. Immediately after 1945, the framework of current planning controls emerged. Plan-making became a key area of planning work in the 1950s. The emphasis of environmental concern was on preserving the communities' intangible `amenity' mainly through protection of the countryside from urban encroachment.

The third phase began in the early 1960s and reached its height in the 1970s; Newby (1990) defines it as post-materialism. While the scarcity of the earth's resources was repeatedly emphasized, the `amenity' issues remained the dominant concern of the British environmental movement. "Scarcity entered to intangible factors such as natural beauty, open space, and pleasing landscapes" (op cit).

Within the arena of planning practice, the emphasis was on accommodating growth. Amenity and environmental quality were still important, but primarily as assets to be safeguarded (Healey, 1993). The various initiatives for the environmental care and management that gathered strength in the 1960s in the face of the expansion of urban development across the countryside did not influence planners' thinking much until the 1970s. The primary concerns of the plans had been growth management, but as the recession set in, there was less growth to manage. At the same time, plans faced a rising agenda of economic and social problems generated by the recession and economic restructuring. As a result and partly because of their post-material character, environmental concerns became vulnerable to the public reordering of priorities. It was a period of major disputes over the scope of planning control between local professionals and central-government officials of the newly created Department of the Environment (DoE). Since then, it has been the government attempts to firstly restrict the scope of planning control to the use and development of land and secondly to keep `environmental protection' issues such as pollution control and rural/nature conservation outside the planning system's direct remit (Grove-White, 1991).

The fourth phase started in the late 1980s with an emphasis on sustainability and the request for sustainable economic development. Attention is now focused on global environmental change and on issues related to resource depletion, and material constraints on rising living standards. As Newby (1990) points out "'ecology' has replaced `amenity' as the focus of public debate".

The shift in environmental concern also contributed in putting development plans back on the agenda after a period of marginalisation in the 1980s. During it, central government used its interpretive power within the discretionary structure of the planning system to experiment with ways of recasting the regulatory regime. `Market conditions' gained priorities in land-allocation policies. Within the newly adopted property-led urban regeneration strategy the quality of the environment became a commodity that could be marketized and traded (Whatmore and Boucher, 1993). Parallel with all this, the existing development plans were progressively sidelined as `out-of-date'.

By 1990, the collapse of the property boom and the adoption of the new environmental agenda put an end to the period of marginalisation for development plans. The introduction of the Planning and Compensation Act, 1991 enhanced the status of the development plan and provided opportunities for local planning authorities to put plans at the centre of decision making over land-use. As plans move to centre stage the most critical new challenge they face is that of sustainable development (Davoudi,1993). The specific focus of environmental sustainability is on biospheric quality and on capacities and thresholds limiting the exploitation and destruction of resources and ecological system, which leads not only to measures to assess the environmental impacts of projects and to require the exploiters, polluters and destroyers to pay to alleviate and compensate for their actions. It also requires attention to be given to control of levels of pollution and amounts of resources use and to ecological capacities and thresholds. This leads back to the causes of pollution and resource use.

12.3 Development plans and sustainable development

The introduction of a plan-led system has led to a new wave of plan-making in the U.K. Throughout the country planners are at different stages of plan preparation to fulfil the government requirement of a full coverage within the next few years. This is at a time when Planning Policy Guidance and Mineral Planning Guidance (PPGs and MPGs) on preparation of plans are also experiencing considerable revisions by the government. The need to update the guidance has been enhanced by the pressure to take the new environmental agenda on board. The new PPG 12, which underlined the government's commitment to an environmental strategy as outlined in `This Common Inheritance' (DoE,1990), pioneered this trend and legitimized the role of land-use planning in sustainable development. It urges planners to "reflect newer environmental concerns such as global warming and the consumption of new renewable resources in the analysis of policies that form part of plan preparation" (D0E, 1992). Since then the DoE has sponsored research on the role of land-use planning in the reduction of transport emissions (DoE/DoT,1993), and the environmental appraisal of development plans. Also, new debates are emerging around the link between transport and development (Breheny, ed 1992; Owens,1992). In minerals' planning, a research report on the role of specification in efficient use of aggregates has recently been published (BRE, 1993).

Nevertheless, as Healey and Shaw (1993) argue, "this infusion of ideas does not yet represent a systematic transformation of the planning agenda, and many ideas are still only tentatively developed". Although the various governmental steps that have been taken in the right direction can contribute to sustainable development, they do not amount to it. In some cases they even portray a superficial understanding of the concept: PPG12, for example, states that "plans must make adequate provision for development and at the same time take account of the need to protect the natural environment" (DoE,1992). Meanwhile the resulting political pressure on planners to "turn the plans green, overnight" has led to some tokenistic approaches such as putting the environment chapter at the beginning of the plan, or providing a list of environmental objectives without having strategies to implement them (Healey and Shaw, 1993).

12.4 Minerals planning

As regards minerals planning, there has been even less attention given to the underlying principles of sustainable development. It seems as if, here, the concerns over global environmental issues have been sidelined by the sheer scale of the environmental impact of mineral operation on the local communities' `amenity'. Within the context of the current economic recession such concerns can be further marginalized by the expediency of political short-termism as it affects job creation.

Minerals can be worked only where they occur and, for reasons of scale, duration and location, this has more impact (or at least more visible impact) on the environment than other development. At the same time, their exploitation makes a major contribution to economic growth. This, in the case of aggregate minerals is supported by a strong `lobby' comprising the mineral and construction industries. Therefore, it is not surprising that the revision of MPG6 concerning aggregates provision has proved one of the most contentious fields of conflict between the emergent environmental consciousness and the established industrial order.

In an attempt to take sustainability on board, the government issued a draft revised MPG6 in January 1993. Whilst there are some `good house-keeping' measures in the document which help to raise environmental efficiency (e.g. increasing the range of supplies, efficiency improvements in minerals use, and substitution out of primary minerals), the revision offers very little towards addressing the sustainability objective with its request for demand management. One of its key requirements is to avoid erosion of the environmental asset base and to maintain environmental resources. The fundamental principle is that expanding supply to meet demand is not sustainable. While government has accepted the environmental challenge to reduce the national demand for energy, water, and travel, it has allowed minerals policy to lag behind.

Draft revised MPG6 offers primarily a demand-led policy approach that hardly questions the depletion of mineral resources. Despite some tentative steps towards the possibility of demand management which have not been followed through the document, the review of the supply and demand equation is dominated by the discussion of supply issues. These include use of lower grade materials where possible, revised standards and specifications, increased use of secondary and recycled materials, enthusiasm for coastal super quarries, and the role of marine dredging. While such measures can raise the environmental efficiency their contribution to sustainability can be questioned, if they are placed in the context of aggregates demand forecasts used by the government. These suggest that the annual consumption of primary aggregates in England and Wales will rise from about 240 million tonnes in 1991 to between 370 and 440 million tonnes in 2011. Secondary materials are thought to provide an additional 10 percent of supply. As Bate (1993) argues, "even to stabilize requirements at 1991 level (similar to the government commitment on co2 emissions), let alone reduce it, would require intervention of one kind or another to avoid the need for between 1,100mt and 1,600mt of aggregates over twenty years". This is in addition to the increase in use of secondary materials already anticipated.

But the government proposals for aggregate supply are far from such interventions. Their key features are "that the demand of the construction industry over the period of 1992-2011 of the order of between 6500mt and 7000mt should be met; that planning authorities should make provisions over that period for some 3000mt of crushed rock and 1600mt of sand and gravel in development plans ; and that the balance would come from a mixture of marine dredged aggregates, secondary and recycled materials and coastal super quarries" (Mabey, Head of the DoE Mineral Division, 1993). Such policy provides no opportunities to change the existing approach to plan-preparation. Planners are still required to prepare plans that meet the demand for minerals in the first place. Yet, in a way which is least damaging to the environment. Thus, the main purpose of the minerals plan is to strike "a balance between the case for mineral extraction and the protection of the environment" (DoE, 1988).

Embedded in this approach is; firstly, the implicit assumption that demands are inevitable and beyond the influence of policy; secondly, that the protection of the environment is limited mainly to the local amenity; and thirdly, that the `balancing' equation does not take into account the capacities and limits of the ecosystem, neither does it appreciate the scarcity and depletion of the mineral resources. A closer examination of the criteria against which mineral operations are judged both in plan preparation and in decision-making over planning applications, will further illustrate the above points.

In preparing minerals plan, planners should, within the plan area, identify either `areas of search' that provide a guide to the industry as to the broad locations where extraction might be permitted; or `preferred areas' where there is a strong accompanying presumption in favour of extraction. In doing this, some planning authorities (eg: Review of the Berkshire Minerals Local Plan, 1991) have adopted a `sieve map' approach which basically means subtracting environmentally constrained areas from mineral bearing areas and identify the ones that pass through as areas of search or preferred areas depending on the degree of precision. This approach is based on geological information and does not take into account issues such as the commercial viability of the workable reserves. A second approach has therefore been adopted (eg: Nottinghamshire MLP Consultation Draft, 1992) particularly in areas where the existing permitted reserves are well over the landbank requirements. Here, the industry initiates the process by putting forward possible future extensions or new sites that will then be considered by the planners as acceptable in principle or otherwise. In both approaches, the criteria against which such judgements are made include:

1. visual impact mainly resulting from fixed plant, stockpiles, overburden and soil orage. This is usually remedied by screening using tree planting and earth ounding;
2. environmental pollution, mainly caused by noise, blast vibration, dust, mud, and water contamination. These are mainly controlled outside the planning arena by various Acts, notably the Environmental Protection Act, 1990;
3. road traffic mainly caused by heavy lorry movements particularly when they ass residential areas;
4. loss of high quality agricultural lands, trees and woodlands, wildlife habitats, istinct landscapes, features of archaeological or ecological importance; and,
5. locational considerations including proximity to residential areas or designated reas of scientific, ecological, aesthetic, or geological importance.

While these provide a reasonably comprehensive list of criteria to protect the local environmental amenity, they do not fulfil the objectives of the new environmentalism, with its concern for global environmental issues and its focus on sustainability. The criteria against which proposals for mineral operations are assessed provide a basis for raising environmental efficiency, that is reducing the environmental impact of each unit of development. But they do not amount to the objective of sustainability, which is to maintain the environmental capacity over time by improving environmental efficiency at least as fast as output growth in order to ensure that environmental quality does not decline below the minimum acceptable level (Jacobs and Stott, 1992). For example, the planning criteria do not include measures for protection of the environmental life support services such as climatic regulation. Neither do they provide measures to limit the exploitation of the non-renewable mineral resources. The visual impact of mineral development particularly in areas of high-quality landscape is one of the top considerations in assessing mineral development. So is proximity to the residential areas. So far, such considerations have had priority over the concern for more critical environmental functions upon which human life depends. Protection of environmental quality in planning arena is still understood as and acted upon mainly in aesthetic and utilitarian terms and not much in biospheric terms. Economic development, on the other hand, is still interpreted in terms of the promotion of economic growth rather than the well-being of all aspects of human life . The following examples from the current mineral planning practice can further illustrate the points raised above.

Resource depletion

Porphyritic andesite or red whinstone is a unique hard rock which is extracted exclusively from a quarry in the north of England. The stone is used as aggregate in road surfacing both in the U.K. and Europe. Its uniqueness is its natural and permanent red colour which makes it ideal for decorative purposes. The red cycle tracks in Holland and Germany, for example, have been constructed using this specific stone.

At its present rate of production, the quarry has a life of about 30 years. Any future extension to the quarry can increase that up to another 30 years. However, any proposals for future extension to the quarry have to meet the criteria listed above. Among them, the visual impact of the operation would have the top priority since the quarry is prominently located in the National Park, and is seen as a threat to the beauty of the surrounding landscapes. But neither the scarcity of the red whinstone nor the fact that it is a finite valuable resource has been (as recorded in the previous planning permissions) and is unlikely to be considered as a constraint on any future quarry extension. In fact, in `balancing' between "the case for mineral extraction and the protection of the environment" the unique quality of the material is a positive point in favour of the former, since it justifies the demand for it.

Global warming

One of the largest whinstone quarries in England is located in the North. It has been in production for over a century. The quarry is located in close proximity to a small village whose residents have been suffering from the environmental disturbances caused by the operation, while appreciating its role in providing male employment. The local community's major complaint is over the lorry traffic. On average, over 200 lorries pass the village road every day.

Any attempt to change this situation is beyond the regulatory power of the planning system, unless via a legal agreement (planning gain) between the mineral planning authority and the operator, or via a new consolidated application when an extension to the quarry is applied for. In either case, changing the lorry routes to bypass the nearby villages can only be achieved at the cost of granting consent to the extension. This means reducing the impact and the nuisance of lorry movement through the village, yet increasing the rate of production and consequently the total number of lorries. In environmental terms, this means compromising the long-term global environmental issues as affected by both depletion of resources (speeding up and widening the exhaustion of the quarry by further extension at a higher rate of production) and emission of co2 (increasing the total number of lorries) in order to achieve immediate local environmental gain (changing the lorry route and maintaining jobs).

Without establishing limits and thresholds for the environmental capacities, in terms of their various functions and the significance of their roles in different aspects of human life, it would be extremely difficult for the planning system to address, let alone to order the priorities of, these issues in a strategic way.

12.5 Conclusion

Since the late 1980s, in Britain, a new wave of public and political enthusiasm has affected both environmental issues and the planning system. At the same time, these issues have experienced a shift from amenity-led to ecology-led approach while the system has seen a change of emphasis from project-based to plan-based approach. As plans move to centre stage the most critical challenge they face is the relationship between environmental quality and economic development. In theory, environmental quality is understood in biospheric as well as aesthetic and utilitarian terms, and economic development is understood not merely as the promotion of growth, nor narrowly as property development, but as the promotion of all aspects of human welfare by adopting the principles of sustainability. In practice, however, such understanding has not yet born fruit. Although development plans have been refurbished as a means of strategically resolving the tension between environmental conservation and economic development, the search for sustainable development strategies has not been given sufficient emphasis. Environmental policies in development plans are still dominated by the issues that are ironically more difficult to grasp, including amenity, aesthetics and concern for the quality of life. But the more tangible issues that at the same time are more critical to the continuation of human life and survival have not yet got onto the planning policy agenda. This is particularly evident in the minerals planning field as reflected in the Regional Guidance for the Northern Region: "mineral planning authorities will need to recognize that in providing for a supply of minerals a balance must be struck between the economic and environmental requirements of the community. The government is committed to the principle of sustainable development which should be taken into account particularly in relation to the landscape, the agricultural, recreational and tourist value of the countryside and the quality of life for local residents" (DoE, 1993).

While the planning system has proved to be an effective device for raising environmental efficiency, its role in providing for sustainabilty objectives has so far been only marginal. No change is likely unless the notion of `balance' in planning system that allows negotiating trade-offs in each particular case is less favoured, and more precise standards and thresholds are established to protect the environmental qualities in a more absolute sense.

References

Bate, R. (1993) "MPG6: The Revision of Guidelines for Aggregate Provision in England andWales", Mineral Planning 55/June

Breheny, M. (eds) (1992), Sustainable Development and Urban Form, Pion, London

(BRE) Building Research Establishment, (1993), Efficient Use of Aggregates and Bulk Construction Material

Davoudi, S. (1993) "Development Plans and Sustainable Development", Paper presented at the vii AESOP Congress, 14-19 July, Poland

Department of the Environment (1993) Draft revised Minerals Planning Guidance 6: Guidelines for Provision of aggregates in England and Wales, HMSO, London

Department of the Environment (1993) Regional Guidance for the Northern Region, HMSO, London

Department of the Environment (1992) Planning Policy Guidance 12: Development Plans and Regional Guidance, HMSO, London

Department of the Environment (1990) This Common Inheritance, HMSO, London

Department of the Environment (1988) Mineral Planning Guidance 1: General Considerations and the Development Plan System, HMSO, London

Department of the Environment/ Department of Transport (1993) Reducing Transport Emission through Planning, HMSO, London

Grove-White, R. (1991) The UK's Environmental Movement and UK Political Economy, Lancaster University

Hajer, M. (1992) "The Politics of Environmental Performance review: Choices in design", in E. Lykke, Environmental Performance Review, Belhaven, London.

Healey, P. and Shaw, T. (1993) Planners, Plans, and Sustainable Development, Regional Studies 27/8, 769-776

Healey, P. (1993) "Discourses of Integration: Making Frameworks, processes and Contents for democratic Urban Planning", Paper presented in Challenges in Urban Management conference, University of Newcastle, 25-27 March

Jacobs, M.and Stott, M. (1993) "Sustainable Development and the Local Economy", Local Economy 7/3 pp. 261-272

Mabey, R. (1993) "MPG6: The Revision of Guidelines for Aggregate Provision in England and Wales", Minerals Planning 55 June

Newby, H. (1990) "Ecology, Amenity, and Society", Town Planning Review 61/1, pp.3-13

Owens, S. (1992) "Energy, Environmental Sustainability and Land-use Planning" in M. Breheny, M. (eds.) Sustainable Development and Urban Form, Pion, London, pp.79-105

Pearce, D., Markandya, A. and Barbier, E. (1989) Blueprint for a Green Economy, Earthscan, London

WCED (1987) Report of the UN World Commission on Environment and Development: Our Common Future, Oxford University Press

Whatmore, S. and Boucher, S. (1993) "Bargaining with nature: the discourse and practice of 'environmental planning gain' ", Transactions of the Institute of British Geographers 18/2, pp. 166-178

Simin Davoudi
Department of Town and County Planning
Claremont Tower
University of Newcastle upon Tyne
United Kingdom

13 OPERATIONALIZING ENVIRONMENTAL CONSIDERATIONS IN THE BRITISH PLANNING SYSTEM

Patsy Healey
Tim Shaw

13.1 Introduction

The land use planning system is now considered by the British government as one of the main arenas which its *Sustainable Development Strategy* is to be achieved (DoE 1993). It provides both a mechanism for articulating local approaches to environmental issues, and a regulatory regime through which measures to address environmental concerns may be pursued. The planning system has traditionally been concerned with issues of environmental protection and quality. There is a strong tendency in contemporary debate to assume that the system, with its distinctive malleability (Healey et al 1988, Brindley et al 1989), can be adjusted to take on board the new agenda. Further, current debates on sustainable urban structures, encapsulated in the concept of "the compact city", allow a return to traditional models of contained settlements and urban restraint strategies.

However, on closer examination, the environmental credentials of the planning system provide an uncertain basis upon which to evolve the local response to environmental issues. Further, the concept of sustainable development, and the UK government's interpretation of it, is itself contested. There are, therefore, major questions to be addressed in operationalizing environmental considerations in the planning system.

These relate to three board but interlinked areas of debate. The first concerns the meaning of environment and environmental sustainability. Within contemporary discussion, there are at one level conflicts over ways of identifying the environmental agenda. Behind these lie deeper epistemological and moral questions about the nature of knowledge, about the relations between people and nature and about our moral responsibilities, to each other and to the natural world. Thus while some, and notably the UK government, seek to contain debate within the vocabulary of contemporary economics and natural science, there are strong pressures for debate to escape into wider "postpositivist" directions. The second area of debate concerns changes in economic and social life as this affects households and firms in places. The task of constructing local planning strategies has had to confront the fragmentation of long established relations between firms and places, new forms of economic activity and changing lifestyles and

D. Borri et al. (eds.), Evaluating Theory-Practice and Urban-Rural Interplay in Planning, 167–192.

new ways of moving around urban and regional space. Exactly how particular spatial patterns for different economic and social relations impinge on each other and with what environmental consequences remains ill-understood. Thirdly, approaches to urban and regional governance are rapidly evolving, linked to the above economic and social changes, and to ideological changes, which both force environmental considerations into the limelight and seek to address them through force environmental considerations into the limelight and seek to address them through regulatory regimes which emphasize new ways of combining economic instruments with administrative procedures (Healey 1993). All these developments indicate that a return to the past is little guide to the way forward in addressing environmental concerns in the planning system.

This paper explores the way environmental considerations are being developed in the planning system in the context of these challenges, focusing in particular on *development plans*. It considers first the environmental debates, in order to arrive at a more precise specification of what "environmental considerations" the system should be expected to take on board. It them briefly summarizes how such conceptions of environment relate to previous approaches in the post-war history of planning. In conclusion, it suggests ways in which the treatment of the environment in the planning system could evolve.

13.2 The planning system, plans and environment

The planning system in Britain consists of a set of procedures for formulating plans and determining applications for development. The objectives and scope of the system are determined by government policy and local interpretation, underpinned by legal review. Very little is specified in law as to the *scope and content* of planning policy, other than that its regulatory focus is on *the use and development of land*. Exactly what this means, and how it could relate to the social and environmental processes which generate land use and development, has always been a matter of controversy. But at a minimum interpretation, the system is centrally concerned with the *location* of development and with its *characteristics*, with what goes where and on what terms.

The role of the *development plan*, in this minimalist interpretation, is to provide the framework within which the criteria for making regulatory decisions can be established. Its formal function has consistently been to provide a strategic and long-term context to decision-making with respect to local environmental change, linking land use allocation and the terms of development to economic, social and environmental considerations, and providing a means to co-ordinate and regulate the flow development projects. In practice, the emphasis on this strategic and coordinative role for plans has fluctuated. The 1980s, in particular, was characterized by a diminished status for plans and strategic planning policies (Thronley 1991). During this period, however, major changes were underway both in forms of development and ideas about the environment.

Both plan-making and environmental issues have been given a new salience by government in Britain since 1989, as a result of the problems resulting from the backlash of the property boom, and the government's own greening strategy[1]. This "return to plans" of the early 1990s has provided an opportunity for these new pressures and concepts to be translated into planning strategies and policies, encouraged by government statements, legislation supporting plans and the new legislative position of the plan in the 1991 *Planning and Compensation Act.* In its recent advice on development plans (DoE 1992), the current environmental agenda appears to be warmly embraced. The professionals too are rapidly absorbing the concepts and terms of the new agenda (Williams 1993; CPOS 1993; Blowers 1993).

These development in the planning field are closely linked to the growth of environmental policy statements and auditing in local government (Raemakers and Wilson 1992). There has also been a recent emphasis on the neglected link between transport development policy and land use planning. However, the planning system has a weak relation to many other areas of sector policy. The tendency has been to create special procedures alongside or within the system for dealing with pollution control, agriculture, mineral extraction and industrial development. Combined with other moves to constrain the remit of the planning system (as in Enterprise Zones, for example; see Thornley 1991), the planning system ability to take a holistic view of local environmental change and its management is limited. Yet, as government now stresses (DoE 1993) it provides a significant regulatory took through which to pursue environmental policies.

13.3 The contemporary environmental agenda: sustainable development and ecological modernisation

The debates about the environmental "problem" strike at the heart of mach of modern philosophy and public policy, forcing not only a challenge to the "drive for growth" embodied in capitalist economic relations, but an explicit confrontation between scientific knowledge and instrumental reasoning, and moral values[3]. In the 1970s, this debate was caste in terms of a choice between *economic growth* and *environmental conservation.* By the 1980s, these debates had become more complex. Both positions came under pressure in the search for some way of "balancing" economic development and environmental conservation, at local, national and global scales. In Britain, the dominant concept describing this goal is currently the principle of *sustainable development.* This has been vigorously embraced by British central government since the late 1980s, and has also been promoted through EC policy debate and initiatives. In parallel, local politics and professional debate in Britain has been absorbing the concept. However, the distance between rhetoric and practice remains large. The concept itself is elastic in definition (see Pearce et al 1989, Jacobs 1991), allowing many interpretations. In the planning field, as noted above it has been used as much to recover long-established concerns -with the long-term, with social impacts and with democratic debate, as to introduce new criteria and perceptions (e.g. Hardy 1990, McLaren 1990, Blowers 1993).

The concept of sustainable development was first articulated in the late 1970s[4] and then publicised in the Brundtland Report (WCED 1987). This argued that economic growth and environmental conservation were not simple opposites, but that forms of economic development could be chosen which would sustain the environmental capacities and relations needed for future generations. The problem has been to give concept an operational meaning. Some argue that the power of the concept lies not only in its breadth, but in its ambiguity and hence potential to command widespread support. As Hajer (1992) argues, the Brundtland conception provided a powerful storyline with a wide influence. However, once the idea has been adopted as a policy principles its actual leverage depends criticality on its operational interpretation.

Within the struggle over meaning, some reject the concept outright *"Deep green"* arguments stress the fundamental challenge of environmental principles to capitalist organisation and technological development. The attempt to make development sustainable is itself unacceptable from this point of view. The challenge is instead to invent new forms of economic organization from the point of view of a "nature first" conception of the relations between people and nature (see Goodin 1992). Contemporary public policy debate in Britain, however, focuses on the different interpretations of the sustainable development concept. Although these remain varied, it is possible to identify broadly two directions. The first represents a *technicist* interpretation of the concept, in which environmental conservation criteria are balanced, or traded-off, against economic development criteria. It focuses on the identification of stocks of environmental assets and their valuation, using the language of environmental economics and instrumental rationality (Pearce et al 1989, Blowers 1992). The more *radical* perception emphasises the constraints on human activity which must be accepted if ecological (or biospheric) systems are to be protected from further life-threatening deterioration and maintained as a resource for future generations. It elides at the margins into the "nature first" ideas outlined above. This focuses policy attention on the *carrying capacity* of ecological systems, the *ecosystemic relations* through which environmental assets are used and the *social relations* through which they are managed. It emphasizes the management of demand within capacity constraints, and uses the language of natural science and moral discourse (Jacobs 1991, Elkins 1986, Hajer 1992, and, in the planning field, Williams 1993) [5].

This wider perspective has encouraged the use of the term "ecological modernisation" in the debate, to encapsulate new directions in the framing of social and economic relations to incorporate contemporary understanding of the significance of ecosystemic or *biospheric* relations (Hajer 1992).

But within these debates too, there are significant divisions between those who focus on the development of technical engineering and economic solutions to mitigating the adverse effects of development and those who argue for restructuring the amount and form of development within "sustainable" ecological, economic and social capacities, understood as relations and not objects. Some plans now refer to limiting development within "sustainable" labour market and housing market "capacities" (see Cambridge Development Plan). Thus the sustainable development discourse[6] within the planning system can be "turned" towards more traditional strategies of urban containment or "growth management".

However, the Brundtland conception arose from the challenge of combining the ecological or biospheric dimensions of environmental concern with the economic and social objectives of societies. It is in particular in the conceptualization of the relations between human social processes and such "natural" or ecological processes that the sustainable development concept is potentially innovative as regards planning discourse. It is therefore helpful to look more closely at these biospheric dimensions.

Jacobs (1991) seeks to provide a definition upon which practical strategies can be developed for avoiding natural catastrophe and promoting intergenerational equity. His definition of the sustainability concept is as follows:

"Sustainability means that the environment should be protected in such a condition and to such a degree that environmental capacities (the ability of the environment to perform its various functions) are maintained over time: at least at levels sufficient to avoid future catastrophe, and at most at levels which give future generations the opportunity to enjoy an equal measure of environmental consumption" (Jacobs 1991, p. 79/80)
This pays down the moral and aesthetic dimensions of the ways we value nature, but has the value that it suggests relatively specific policy criteria. It also emphasizes the concept of carrying capacity with implications of limits, linked to relations, rather than stocks. Jacobs the defines environmental capacities and functions in "biospheric" terms. The "environmental crisis", Jacobs argues, has arisen because human activities are reducing the "abundance" of the biosphere upon which human life (including the prospect of improvements in economic welfare) depends. This biosphere performs three principal functions:

1. it provides *resources*
   - non-renewable (e.g.: fossil fuels)
   - renewable, so long as critical thresholds are not exceeded (e.g.: clean, air, water)
2. it assimilates *waste products*
3. it provides various *environmental services*
   - amenities (e.g.: recreation space, space for aesthetic enjoyment)
   - life support services (e.g.: stores of genetic diversity) (Jacobs 1991 p. 3-5)

This list generates an agenda of impacts to be considered, whether the objective is a balancing trade-off or demand-management within capacity constraints: it parallels many of the agendas of environmental issues found in contemporary policy reports (e.g. DoE 1992, CPOs 1993)
Developed in this way, the concept of sustainable development offers a new approach to the traditional planning challenge of relating the economic, social and physical dimensions of human existence. It emphasises the processes of environmental management, and the relation of specific actions to ecological relations and capacities. It introduces new criteria within the planning calculus, grounded in both natural science and non-utilitarian conceptions of welfare. It challenges notions of planning as an exercise in merely *balancing conflicting interests* understood in simple trade-off terms, and emphasises ecological relations, environmental limits and capacities. It thus forces consideration of projects in their environmental, social and economic relations, rather than as discrete buildings-on-sites.

13.4 Environmental sustainability and planning debate

There is currently an explosion of discussion in the planning field seeking to identify the implications of incorporating the new environmentalism in land use planning[7]. The EC *Green Book on the Urban Environment* captures the debate in the planning community of many European countries with its mixture of long-standing concerns for the quality of the built environment and the new biospheric agenda (EEC 1990). While several European countries used the environmental agenda to re-emphasize traditional concerns for the conservation of the built and natural environment, others have been substantially ahead of Britain operationalizing biospheric environmental policies in spatial planning. In Britain, discussion was slow to move from a traditional agenda, but rapidly shifted in the early 1990s. An indication of the speed of development of ideas can be seen by contrasting the planning section on *This Common Inheritance*, which barely touches on the new agenda[8], and *Planning Policy Guidance* 12 (DoE 1992) which has a broader and much more informed approach. The pace of innovation has speeded up in planning practice. Rydin (1993) identifies what are in effect tentative steps in incorporating new policy criteria. But by 1993, the County Planning Officers were using the language of *carrying capacity* (CPOS 1993, Williams 1993) and Government Ministers were firmly asserting the need to limit CO_2 emissions and car use within regionally established thresholds[9].

In planning debate in Britain, discussion focused in the early 1990s on two issues; firstly the relation between energy conservation and pollution reduction through acting on the relation between land use and transport, (OECD 1993, DoE/DTp 1993; Breheny et al 1992), and more generally the relation between urban form and environmental sustainability. (Breheny 1992, Breheny et al 1992). This reflects not only the significance of energy use in consuming non-renewable resources and in local and global pollution. The environmental dimensions of transport may also be linked to concerns with public finance, and with the costs of continuing provision for road traffic demand.

However, the environmental agenda implied in Jacobs' approach to the environment's biospheric functions suggests a broader approach to identifying the way the planning system intersects with environmental issues. Such a broader approach is evident in the recent work of several scholars seeking to elucidate the intersection (see Owens 1992, Rydin 1992). How far particular policies will achieve environmentally sustainable objectives depends on the relations of local natural ecosystems and on the social and institutional relations through which land use planning actions are taken. It is often argued that land use planning action can have a significant impact on achieving environmental objectives, but usually only if combined with strategies for other sectors at national, regional and local level (Nijkamp et al 1990, DoE/DTp 1993).

Most recent reports which seek to define the range of environmental issues which should now be considered within the planning arena tend to focus on substantive topics (such as waste recycling, biodiversity, environmental quality etc., e.g. CPOS 1993). An alternative approach is to focus on types of action which should be pursued through the planning system. Figure 1 seeks to do this, drawing on Jacobs' list of biospheric functions outlined above.

Several of these areas of action are long-established activities for the British planning system, notably the first, second, fourth and fifth. Further, it is these tasks which the system has been in the past most effective and achieving (Hall et al 1973, Healey et al 1988, Pearce 1992). This might suggest that regulatory action through the planning system is a particularly effective mechanism within which the objectives of sustainable development can be accommodated[10].

Figure 1

The areas of action for land use planning systems in environmentally sustainable strategies
1. Conservation **- of sites** **- of environmental qualities** **- of building qualities**
2. Location of development **- to conserve energy/reduce pollution** **- to promote pollution reduction** **- to provide amenities** **- to promote biodiversity** **- to limit development within capacity thresholds**
3. Define capacity "areas" within which standards/targets to be met
4. Identity sites for environmentally desirable development
5. Promote environment-enhancing qualities in all development
6. Mitigate the adverse distributional effects of environmentally beneficial strategies

If the above were a correct assumption, then the innovative attention for contemporary planners and planning policy makers should focus on the two less familiar areas of action in Figure 1, introducing pollution control areas and emphasising mitigating the environmental damage. To some extent, this is already happening. There is now considerable discussion of the need to link "environmental permitting" and the new Integrated Pollution Control machinery more closely with the planning system[11]. Meanwhile, there is a strengthening tendency to consider all development proposals in terms of their economic social and tendency to consider all development proposals in terms of their economic, social and environmental impacts, and to negotiate measures to alleviate or compensate for those impacts where a project is otherwise in line with established planning policy (Healey et al 1992; Lichfield 1989, 1992; Cowall 1993). The development of impact identification in effect reconstitutes the approach to "balancing" in the planning system, in line with the less radical version of ecological modernisation (see CPOS 1993, and Williams 1993). The influence of EC legislation and the requirement for its adoption within British law has been a major factor in promoting these ideas.

This suggests that the early 1990s might be an auspicious political movement for a substantial general reformulation in thinking about both the form and content of the planning system. However, on closer examination, the planning system's record in achieving environmental objectives in the past provides only limited reassurance about its capacity to meet the demands of the strategies of sustainable development. The system's powers have regularly been constrained, to protect the interests of particular business sectors. The very flexibility of the system's form allows the distortion of broad objectives through implementation (Healey et al 1988).

The history of the treatment of environment in development plans since the 1940s serves to illustrate this point. This history not only charts an evolving range of conceptions of the environment. It also shows how dominant economic and materialist conceptions have persistently marginalized alternative conceptions and their leverage on plan content.

13.5 The treatment of "environment" in development plans: 1940s - 1990s

Phases in environmental understanding

Newby (1990) argues that there have been four phases in the treatment of environmental concerns in the planning system, -an early period when the emphasis was on preservation of a pre-industrial past, an inter-war period when the emphasis was on combining preservation from development and regulation of development to enhance and safeguard amenities, and from, the early 1960s, a renewed realisation of the scarcity of environmental assets confronted with the pressures of local and global economic and demographic growth. He then claims that by the 1990s, " 'ecology' has replaced 'amenity' as the focus of public debate" (p.8). Newby's phases in effect represent four different discourses within which "environment" is given meaning. What more (1993) develops this analysis more explicitly in her work on the environmental discourses in planning system, arguing that a *conservation* narrative was dominant in the consideration of environmental issues until the1980s[12].

The analysis of policy discourses is always a complex matter as many discursive strands are likely to co-exist. One challenge of development plan preparation is to select from and adjust to what is often a competition between discourses. This is particularly so in the environmental field, within which multiple perspectives co-exist and compete for attention (O'Riordan 1980, Harvey 1989). The text of development plans thus provides a useful research focus through which to explore the way the 'environment' has been understood. In a recent study, we have looked at a range of plans and related reports from the 1940s-1950s, to the 1990s (Healey and Shaw 1993). From this a more nuanced history of the environmental discourses of the planning system emerges than the accounts offered by Newby and Whatmore. We suggest that five strands can be identified:

1. a welfarist-utilitarianism, combined with a moral landscape aesthetic (1940s on)
2. growth management, servicing and containing growth while conserving open land (1960s on)
3. active environmental care and management (1970s on)

4. a marketised utilitarianism, combined with conservation of nationally important heritage (1980s on)
5. sustainable development (1990s)

These strands do not neatly succeed each other. Rather, they have co-existed, older ideas persisting as newer strands are brought into the play. Ideas which are now emphasised have been advocated before. Policy proposals have been sidelined because they were at odds not merely with dominant economic and political interests, but dominant conceptions of nature-society relations, and of the relation between state, economy and society. Throughout the period, it is the modernist rationalist discourse of economic growth, instrumental rationality, and the ability of the human species to control and manage nature, which has won out over other conceptions.

The 1940s/1950s: A utilitarian aesthetic

In this period, the strategic ideas developing through the inter-war period with respect ot economic development and rural character were absorbed into a comprehensive land use planning framework. Key influences were the ideas of leading planners, notably Patrick Abercrombie and Thomas Sharp. Both wrote manifests about planning (Abercrombie 1933), and prepared a number of regional and city plans. Both argued for a clear separation of town and country on moral and aesthetic grounds[13].

Such ideas were followed through into Abercrombie's influential Greater London Plan (1944). Here, Abercrombie combined an aesthetic emphasis with a focus on functional organisation. The natural environment is primarily treated as a *backcloth*, which provides resources for exploitation (for mineral extraction and farming), opportunities for healthy recreation, and landscapes for "the visual solace fo man" (p.3). Farming was seen as inherent in "the normal countryside" (par 65, p. 18), but mineral workings were the source of disfigurement. Planning was presented as having a stewardship role with respect to the natural environment, producing a valued inheritance for the future. This precursor of the sustainability principle derives from the long-established concept of the landowner as steward of the countryside which had such an influence on rural development policy until very recently. Nevertheless, overall, the emphasis in the plan is on the environment as a collection of objects to be preserved and amenities to be enhanced. The concern was thus with Jacobs' *environmental services* functions of the biosphere, but conceived in a moral aesthetic of idealised relations between people and landscape.

This *preservationist* priority and also to accommodate the demand for development. This tension was focused around the notion of the countryside as a resource, for agricultural production, as an aesthetic landscape to be conserved and as a recreational resource. The concern for agricultural production was particularly strong. The debate over development needs versus land resources for agriculture was underpinned by powerful national industrial and farming interests. A different concern was embodied in the campaign to open up the countryside ot the mass of the population, a movement strongly supported by the Labour Party and the Unions in the inter-war period. This issue co-existed uncomfortably with the interests of those seeking to conserve the traditional rural landscape, as well as the national priority to promote agricultural production.

The preoccupation of the plans of the period was with quality of life and quality of the environment. The needs of industry and economic activity are discussed only tangentially. This did not imply a prioritising of social and environmental considerations over economic ones. Rather, it reflected the assumption that economic growth was able to generate its own dynamic. The role of the planning system was to accommodate growth, while at the same time improving social and environmental conditions locally. Local planning policy supported industrial activity by allocating space, by ensuring adequate physical infrastructure, and by managing spatial arrangements to ensure workers could get to factories.

The planning tradition reflected in these plans embodies a peculiarly British marriage between economic modernisation and a romantic nostalgia for a particular ideal of rural life and landscape, a *utilitarian aesthetic*. It illustrates an awareness of the environmental costs of economic activity and promotes efforts to ameliorate them. In general, however, the planning system's role in economic development was limited to providing space for industry and ensuring an available labour supply with reasonable living standards. Environmental considerations only limited economic activity "at the margin". This was partly because of lack of knowledge of environmental processes and pollutants. Noise, dirt and smell were obvious. Other environmental damage was not. Nor was the car seen as a threat. But a second reason why environmental dangers were marginalised was the dominant functional and materialist emphasis. By the later 1940s, the power of this dominant emphasis was such that Abercrombie and Sharp's moral aesthetic was increasingly sidelined.

The 1950s saw a further emasculation of the early conceptions of environment, as the plan-making under the Town and *Country Planning* Act 1947 got underway. The legislation was itself more limited than many protagonists had hoped for (Cullingworth 1975). Thus, the environment was primarily seen in terms of material resources and aesthetic qualities, as a resource "container" and a "backcloth", to be safeguarded for economic and social enjoyment. This discourse was pursued through a regulatory regime which privileged economic activity through the protection of the "national interest". The result was a progressive narrowing of the environmental agenda in the planning system.

The 1960s: Managing growth.

By the 1960s, the emphasis had shifted to accommodating growth and further modernising town centres. This brought not only further new town projects, but renewed interest in strategic planning and development plans, after relative disinterest at central government level during the 1950s. The result was a series of regional studies, new town proposals and new legislation on development plans.

The emphasis in these plans and proposals was primarly on development and the physical environment. Liverpool City council's *City Planning Policy Report* 1964 for example made little reference to the natural environment, except as open space, which is described as "the setting for physical recreation" (p. 17). The South Hampshire Study 1966, commissioned from Colin Buchanan and Partner s by the Ministry of Housing and Local Government and the Hampshire local authorities, gives more attention to natural resources. These are listed as gravel, sand, agriculture, forestry, fishing and "natural history", this latter term referring to wildlife and their habitats. But the central preoccupation was the dilemma of accommodating growth, in a political situation where local people increasingly sought to limit urban expansion to protect their landscapes, their property assets and their lifestyles; i.e. growth accommodation. Other plans of this period for growth zones and new town proposals reflect an emphasis on the built environment. An emphasis on environment as setting pervades the *Plan for Milton Keynes*, produced in 1970 by consultants (Llewelyn Davies, Weekes, Forestier-Walker and Bor). This sought to maximize access and to provide for car-based expansion, but key structural features of the plan were the landscaped parkways and the design of neighbourhoods intended to be "safe, quiet and pollution free" (p 39).

These plans and studies were the product of a planning culture dominated by architects and engineers. Environment was landscape, natural history and a setting for recreational activity and enjoyment. It was also seen as a set of environmental services (facilities for recreation and for aesthetic enjoyment), and as a store of resources for exploitation.

However, during the 1960s, concern for the countryside, its appearance and its management, reappeared in policy debate[14]. This was linked to the scale of the pressure for development in the countryside during the 1960s, and the growing understanding of the natural science of the environment. Thus the treatment of the countryside in plans, shifted slowly towards the contemporary environmental agenda. However, the approach was still rooted in a conception of environmental quality for people to enjoy. There is no consideration of environmental capacities or limits, merely of adverse impacts that will need to be dealt with (as in restoration of mineral workings, or measures of reduce river pollution). There is little sense of the idea of environmental stewardship evident in planning thinking in the 1940s.

To conclude, the preoccupations of this period were with growth accommodation and management. The natural environment was treated primarily in functional terms, as resource and amenity, with a residual aesthetic as "setting". An alternative resource husbandry approach was sidelined until late in the decade by the dominant urban structure and design preoccupations of the influential planning consultancies of the period. By this time, the new environmental agenda was being articulated more forcefully, and the intellectual driving force in the planning profession was moving from architecture to regional geography and policy science.

1970s: Active environmental care

This shift was well-expressed in the sub-regional and regional plans of the period. For example, the Strategic Plan for the South East 1970 drew its intellectual inspiration from regional geography and represented a very different approach to analysing development requirements and how to satisfy them. Yet it made few innovations in its treatment of environmental issues, focusing primarily on the countryside. This was presented as a land reserve for development. It is a source of essential materials, a location of much of the nation's natural heritage, and a resource for recreation. The plan also gave some attention to the changing farm economy and its impact on landscape, and the problem of competition between mineral extraction development and landscape quality. Although there is the hint of a "sustainability" objective in the concern to limit development within regional water capacity constraints, the plan's general approach to the environment is as a resource for the benefit of "the South East as a whole" (par 4.42).

Central government throughout the 1970s sought to contain the innovation that was occurring as counties prepared their structure plans. Local authorities were exhorted authorities to be selective and focus on key issues. Employment, housing and transport were intended to be the primary concerns of structure plans. Other key issues could include the conservation of character, provision for recreation and tourism, and the location and scale of land reclamation (DoE 1974). Agriculture and pollution recuction measures could be included in plans (DoE 1977), and by 1979 structure planners were encouraged to assess the impact of other policies on agriculture (DoE 1979).

In parallel with these specifications of the form and content of development plans, the 1970s saw both the development of the work of the Countryside Commission and a series of inquiries into countryside issues[15]. Countryside *management* hed by 1979 become a major preoccupation for local planning authorities.

The *Greater London Development Plan* 1976, the first major plan of the period, contained little innovation with respect to environmental issues. However, the structure plans produced towards the end of the 1970s reflected the growing interest in countryside managment, as well as changes in professional thought and popular concern in response to the growing leverage of the environmental agenda[16]. Authorities were often engaged in environmental enhancement work, promoted by bodies such as the Countryside Commission, and by the availability of derelict land grant for a wide range of environmental improvements. This encouraged authorities to pay more attention to environmental care as well as protection, and to *active environmental management*. Plans for both the metropolitan areas and the counties began to emphasize urban fringe management, improving water quality, limiting agricultural damage and control mineral exploitation. Yet these plans continue to treat the environment as available for exploitation. The main concern was to reduce degradation and enhance productive and amenity qualities. There is still non conception of limited *capacities* or thresholds.

All these plans stray beyond actins which could be pursued through land use planning powers. They also challenged government policy. Yet they offered a strategic agenda for the coordination of a range of different types of action. Central government's reaction was to seek to limit the lanning agenda, both with respect to environmental management and social issues. The justification was partly a technical argument about the way actions under different powers should be cross-referenced. But it also reflected a political argument about the role of local government in local environmental management.

The 1970s thus saw a shift in planning discourse about "environment", from a conception as a resource for exploitation and as a setting, to rural resource managment. This still embodied the mixture of utilitarian/functional concerns and aesthetic considerations inherited from earlier periods, but the potential conflicts between these were recognised much more clearly; hence the emphasis on managing multiple activities in the countryside. The major change was that the countryside was no longer merely seen as a setting or backcloth to be conserved, but in terms of *natural systems*, to be managed, to safeguard their economic and amenity value.

However, such countryside management challenged the bastions of rural development power, the landowners and farmers, (the remit of the Ministry of Agriculture), as well as the minerals industry. In this context, central government's resistance to active environmental management in development plans is one reflection of the struggle for power between these bastions and a new local countryside politics developing as middle class urbanites moved out into rural environments. There was less tension between environmental conservation and management and economic development in the 1970s, because developers were typically still working out the land allocations made available through the growth accommodation strategies of the 1960s, the slump of teh mid-1970s having slowed down the pace of development[17]. The 1980s however brought the conflict between environmental conservation and management and development to the fore, both through the introduction of neo-liberal political philosophy into central governement thinking and through market pressures. These came head on against a maturing popular understanding of, and concern about, environmental issues.

The 1980s: Marketised utilitarianism and heritage

The widening environmental and countryside agenda of the 1970s is reflected in the content of many of local plans of the early 1980s, and in comprehensive government advice on the content of plans produced in 1984 (DoE 1984). This circular was in retrospect a swansong for the professional agenda of planning issues which had built up through the 1970s. It came up against a new impetus from central government, driven by ministers and some civil servants, to simplify and reduce the scale and scope of planning regulation. The two specific targets were to speed up development control decision making and to demote the status of development plans. The focus of debate was on balancing *development and conservation*.

In practice this meant an approach to land allocation, particularly housing land, which took on board commercial criteria in determining amounts and locations of development, and which treated the environment as landscape assets to be separated off and protected from the threat of development. The result was to weaken the restraint policies developed in the 1970s and allow substantial possibility for urban expansion beyond established urban areas (Rydin 1993; Thornley 1991).

This impetus culminated in Circular 14/85 (DoE 1985). This emphasised the "presumption in favour of development" which in central government's view should be the priority principle pursued through the planning system. Development projects were only to be refused if they caused *"demonstrable harm to interests of acknowledged importance"* (p.1). Development plans were relegated in status to a material consideration on a par with any other. The result was what came to be known as an appeal-led rather than a plan-led planning system, with policy issues debated on a project-by-project basis, rather than in relation to a planning framework. This was a difficult arena within which to discuss the complex relations between the new environmental agenda, and socio-economic priorities.

Government policy thus both narrowed the planning agenda, and shifted it to absorb criteria related to property market considerations (Brindley et al 1989, Thornley 1991). As Whatmore (1993) notes, this encouraged a conception of sites, buildings and environmental qualities as *commodities*, the generation and trading of which was to be regulated through the planning system. This emphasis on the environment as tradable assets (qualities and facilities) was combined with a narrow conception of conservation as heritage landscapes and wildlife sites. Whatmore refers to this as a conservation narrative, although in its 1980s version it represents a narrower conception than in the past shorn of concerns with either stewardship or active environmental care.

Local authorities were beginning to innovate however. As in so many other respects, the Greater London Council's proposed revisions to the Greater London Development Plan, firmly introduced a new agenda, and provided a useful source for other authorities seeking to "green" their development plans in the later 1980s. The environmental agenda was given further encouragement by the introduction of requirements of environmental impact assessment on major projects, introduced in 1986 in order to comply with EC policy.

But the thresholds for such requirements were set high, and EIA has so far had only a limited impact on the flow of development control decision making and planning policy. (see Lykke 1992).

An important locus for innovation lay outside the planning system, in the preparation of *environmental statements* or *environmental charters*. These were often inspired by local politicians, and developed by officers in environmental health as well as planning departments. Kirklees Metropolitan Borough Council was pioneer in this respect, producing, in collaboration with Friends of the Earth, a *Charter for the Environment* (KMBC/FoE no date). The London Borough of Sutton was another pioneer, producing a series of annual Environmental Statements from the mid-1980s. By the end of the decade, both Friends of the Earth (1989) and the Local Government Management Board had produced guidance to local authorities on the content of green charters (Bosworth 1993).

This emerging environmental agenda was not reflected significantly in development plans until the early 1990s. The planning system was during still constrained in its response to the new agenda by the attitude of central government, and by the slow percolation of the new agenda into the planning profession[18].One impetus for this derived from the economic development arena, where the importance of environmental quality in supporting economic development initiatives was increasingly stressed. The qualities involved primarily related to visual attributes of status labelling, along with the provision of infrastructure and management services. However, the result was that environmental quality and economic development were increasingly presented in plans not as opposites, as in government thinking in the early 1980s, but as a "positive sum" relation.

By 1990, however, the treatment of "environment" in plans began to change substantially, reflecting both changes in professional understanding, and popular pressure. The adverse development and congestion consequences of the economic booms and slump coupled with the strength of generalised public support for the new environmental agenda, helped to lay the political foundations for a dramatic U-turn in government policy on both the environmental agenda and the role and content of development plans[19]. There was symbolised by the publication of *This Common Inheritance* in 1990 (DoE 1990), the result of a government-wide exercise intended to show how government was taking on the agenda of environmentally sustainable development in all areas of its work.

The 1980s thus saw dramatic swings in government attitudes to planning and to environmental questions. These first impeded and then accelerated the development of appropriate responses to the new environmental agenda by local authorities and the planning profession. The development of approaches to active environmental management was curtailed at the start of the decade. Instead, a narrow utilitarianism was promoted which treated environment as a collection of tradable assets or commodities, as Whatmore (1993) argue. During the mid 1980s, the value of such "environmental quality" assets in relation to economic development strategy was increasingly appreciated. It was only late in the decade that the environmental sustainability debate was recognised, and, with the exception of a few pioneers, the biospheric and resource conservation dimensions of the environmental agenda were largely neglected unless part of traditional agendas.

One explanation for these swings in policy attention and discursive formulation can be found in the hope and ultimate failure of the strategy of promoting a market-led approach to the amount, location and form of development (Thornley 1991, Healey 1992). This not only served to undermine property development makets in themselves. It also activated a wide-ranging political backclash as the cost of such an approach to development management came to be widely appreciated. But beyond this narrow concern, the "environmental turn" in British policy debate in the late 1980s reflects a belated appreciation among policy elites of the supra-national and global dimensions of the environmental debate (Hajer 1993)[20]. It was therefore not until the early 1990s that the operationalization of the environmental sustainability agenda within the planning system really begins.

The 1990s: Sustainable Development and Carrying Capacity: The Operational Struggle.

1990 marked a major shift in the climate of thinking in Britain with respect to environmental policy. With *Our Common Inheritance*, government committed itself to the new environmental agenda, and began to translate this into its implications for all areas of government policy. With respect to planning, these efforts wer complemented by EC interest in the urban environment, with the publication of the *Green Book on the Urban Environment* (CEC 1990). By 1992, national planning advice on plans strongly emphasised environmental issues (DoE 1990). By 1992, national planning advice on plans strongly emphasised environmental issues (DoE 1992), wiht further development in guidance on land use and transport in 1993 (DoE 1993). This encouraged the innovative local authorities to develop their ideas more vigorously, and pushed all authorities to re-think their approach.

Initially, the debate within the planning system was over the content of the agenda. This implied adding further topics to the range already included in the planning remit. Given the malleabiltity of the system, these were relatively easy to absorb. By 1993, however, the more fundamental debate between technicist and radical conceptions of environmental sustainability and ecological modernisation were affecting discussion. The vocabulary of limits and capacities and of demand managment was given increasing prominence, and contrasted with balancing and trade-off approaches. (Williams 1993, CPOS 1993). Both approaches encouraged consideration of the impacts of development, and whether and how to mitigate the impacts[21]. They also opened up old debates on spatial strategy and urban structure.

National government policy with respect ot the planning system was articulated most clearly in the revised *Planning Policy Guidance* (PPG12) with respect to development plans (DoE 1992). This includes a spceial section on *Plans and the Environment*. It emphasises the need for a shift of attention:

"Local planning authorities should take accoutn of the environment in the widest sense in plan preparation. They are familiar with the "traditional" issues of Green Belt, concern for landscape quality and nature conservation, the built heritage and conservation areas. They are familiar too with pollution control planning for healthier cities. The challenge is to ensurethat never environmental concerns, such as global warming and the consumption of non-renewable resources, are also reflected in the analysis of policies that form part of plan preparation" (para 6.3).

In developing its ideas, PPG 12 gives considerable attention to the role of planning in energy conservation, and specifically in reducing CO_2 emissions. This leads to dicussion of ways of limiting car travel, propting development near public transport and othe measures. However, the conception underlying this discussion is clearly on of trade-offs[22].

The notion of environmental limits and of carrying capacity barely appears in government statements until 1993, when Minister Michael Howard used the concept of *demand managment*, with its implications of managing demand within capacity constraint when launching a draft Planning Policy Guidance Note on Transport[23].

This suggests a rapid evolution towards the more radical conceptions of environmental sustainability. Yet at the same time, there are vigorous efforts to contain the debate. The recent DoE statement on the UK *strategy of sustainable Development* (DoE 1993) pulls back to the vocabulary of stocks and trade-off, and emphasizes the contribution of "good science". DoE regional offices have meanwhile been amending policies in plans which suggest a "presumption against development" on environment grounds. This is seen to be out of line with government policy to promote development. What this suggests is that the DoE is seeking both to take control of the environmental agenda and define it in technicist terms.

However, local authorities have now taken the innovatory initiative, often with strong local political support. The work on green charters, statements and audits in local authorities in the late 1980s provided a foundation for defining stocks, targets and capacities. In reviewing local authority progress in relation to the treatment of environmental issues generally, Jacobs (1992) suggests authorities have developed their policies through three phases. The first treats environmental issues inisolation, within specific policy areas; the second develops a holistic approach, looks acros the range of local authority activities, and seeks to develop an integrated approach to policy and action. Many local authorities are at this stage. Phase three invoves a movement to the adoption of sustainable deveopment principles, with a choice between the weak, or balancing, version, and a strong, or constraint-oriented, version. A few examples of plans are now appearing which illustrate tentative moves towards Jacobs' phase three (see Newcastle UDP, Cambridge DP, CPOs 1993, Kent Structure Plan).

Thus the environmental discourse expressed within development plans and government guidance of the early 1990s moved decisively away from the narrow utilitarianism of the 1980s which set development and environment in opposition to each other. The debate is now about the meaning of sustainability, with the terminology of balance and trade-off competing with that of limits, constraints and demand management as government and local authorities struggle to understand the conceptual and operational options.

The environmental policy momentum in Britain is now very much in the administrative and professional arena. The implications of this for the power relations supporting the "greening" of British government policy, and specifically planning policy, are not yet clear. As the operational issues are more clearly understood, their inherent difficulty, both in terms of local variability,a and with respect to real choices over short term economic interest and long-term environmental quality, is likely to become much clearer (Owens 1993). Many commentators fear that the "professionalisation"of environmental discourse is serving to divide the new agenda from the popular support it achieved in the 1980s, and thus from the political base to maintain its leverage when the hard choices, both locally and nationally, have to made (O'Riordan 1993, Blowers 1993, Owens 1993, Hajer 1992). More fundamentally, there are strong signs of an attempt by central government and its advisers to contain the environmental agenda within the narrow confines of economic technique and positivistic natural science, denying the moral and epistemological complexity of the issues at stake (Hajer 1992, Harvey 1989).

13.6 Environment, economy and planning

Three approaches to the treatment of environment in development plans run thruogh the postwar period and prefiqure the current debate over the meaning of sustainable development. The first emphasises the environment as *functional resource*, a reserve of non-renewable resources and amenities for human enjoyment. The environmental concern is with their *conservation*. During the 1980s, the conception of a "reserve" was increasingly transformed into a notion of tradable assets or commodities, to be priced, using the techniques of environmental economics (Whatmore 1993). This provides a strong foundation for the interpretation of sustainable development in terms of a stock of assets although the specifics of the discourse have had little impact on the vocabulary and methods of development plan policy specification.

The second strand emphaisises the moral and aesthetic notion of the environment as backcloth or *setting*. The moral dimension of this conception was very clear in the thinking of the pionner planners, but was soon sidelined into a narrow view of conservation. What evolved instead during the 1960s and into the 1970s was renewed interest in active *stewardship* of the natural environment. This provides support for the *asset stock* conception of sustainable development, in the sense that the stock needs to be improved and its deterioration actively prevented. But it also provides support for the more radical interpretation of sustainable development, in that it focuses attention on stewardship as a moral duty.

The third strand is less clearly articulated in early environmental debates in the planning system. It is most clearly seen in the discussions in the North West plans on how to accommodate expanding agricultre, industrial development and mineral extraction, while also imoroving the quality of urban and rural life. Constraints on economic and socila development were deemed necessary to conserve agricultural life and landscape, and to bring air and water pollution within new quality thresholds. The sieve map techniques used to identify where development should and should not go also echoes a conception of environmental constraint. What is new is the uderstanding of the ecological dimensions of such constraints, and their variable local, regional and global impacts, and the recognition that attention needs to be given not merely to stocks and qualities, but to environmental and social *relations*, the destruction of which leads to degradation. Thus the radical vocabulary of environmental limits, thresholds and demand managment has precursors in planning debate.

Yet despite the continuity in these trands of environemtal debate, there can be no doubt that the postwar history of the planning system has seen the dominance of economic over environmental considerations, just as a narrow environmental conservation allied with economic emphases allowed the sidelining of social distribution concerns (Hall et al 1973, Ambrose 1986). The economic dominance has been achieved in various ways. At the level of policy discourse, the dominant emphasis in British public policy has been utilitarian in form, focuses around instrumental rationality. This was strongly emphasized in the 1980s (Drucker et al 1985) and continues into the 1990s. Within planning debate, conceptions of the moral value of nature, of environmental stewardship and of preserving an inheritance for future generations have been steadily sidelined. Even the "traditional" view of environmental conservation as the management of landed estates has given way to more financially dirven conceptions of economic priorities, with the associated emphasis on assets, asset trading and trade-offs. Within the palnning system, the tradition of administrative discretion rather than legal ruel, and the cultivation of flexibility in administrative guidance, has allowed economic interests to be subtly prioritised, in administrative interpretation as well as in more public disputes over plan content and development control.

13.7 "Entrenching" environmental sustainability conceptions within the planning system

The challenge for the new environmental agenda is therefore not simply one of developing appropriate conceptions, policy instruments and skills in local operationalisation. It is a political challenge for real leverage over economic discourse - at the level of policy and practice. Only if this happens will the sustainability objective of a beneficent relation between economic development and global/local environmental quality be achieved. While this is so whichever of the two interpretations of sustainability are adopted, the current governmental preference for the conception of balnces and trade-offs not onlysits more comfortably with economic priorities. It is also more easily subverted by economic growth imperatives in that environmental limits to trade-offs are not set.

Within environmental debate in the planning system, operational ideas are developing apace (Breheny et al 1992, Williams 1993, CPOS 1993, Owens 1993, Jacobs 1992, Blowers et al 1993). One effect of these debates is to discover the value of "traditional" planning policies. "Old friends" are being refurbished and reinterpreted in the context of the sustainabilkity debate, for example conceptions of balancing and weighting interests and impacts, the long-standing idea of contained development and the "compact city"; the value of public transport and the parallel between the urban structure idea of "decentralised concentration" (Owens 1991, Breheny et al 1992) and Howard's notion of garden cities. This gives support to those planning officers and civil servants who want to argue that past planning policies turn out to have been quite environmentally friendly. By implication, more of the same could be a sufficient response.

Yet this ingnores the scale of the challenge outlined above. Traditional planning methodology has typically been judgemental rather than calculative in form, drawing more on legal forms of argument than economic ones. As such, there are major problems incorporating the language of trade-off and balance in a calculative form. The increasing interest in identifying the impacts of development in order to address more systematically the dimensions and values of trade-off judgemnet require more systematisation and precision if the notion of an environmental stock is to be operationalised. In the present context, with the dominance in government policy of instrumental rationality and economistic coneptions of public policy practice, it is not enough to point out that converting environmental considerations of questions of stocks and trade-offs is a narrow and reductivist approach to the issues at stake. It is necessary to illustrate these limitations by attempting to work out what such an approach would involve. Thus,a as a critical enterprise, the efforts of those developing a calculative approach to balancing environmental and economic considerations should be welcomed, as Jacobs (1991) argues.

But developing the more radical conception of environmentally sustainable strategy requires attention at both the level of technique and the level of value. As regards technique, more work is needed to operationalize conceptions of limits and carrying capacities understood in terms of socila environmental relations, and of demand managment within these capacities All the work currently underway in local authorities on enviromental audits target setting, developing monitoring indicators and on the environmental impacts of plans should helpful in this respect[24]. Several planning authorities are now working on these issues, whihc should make it easier to "tell the difference", -between a plan whihc has real potential for environmental leverage on economic development and one whihc does not; and between one which moves beyond a balancing conception fo sustainable development, to one based on a relational approach to enviromnetal capacities.

The debates are not, however, solely a question of technical calculation . How we conceptualize both economy and environment is inextricably bound up with conceptions of value, with respect to our relations to nature, to each other, and to economic life. Contemporary environmental economics argues that value can be indentified in terms of individual preferences for the conservation of particular assets and stocks. The limits of such instrumental rationality in dealing with values in public policy is now widely understood. It is acknowledged in legal argument, which focuses on the reasoning and reasonableness of a judgement. This allows moral principle and political ideology to enter into the decision arena explicity.

Through processes of argumentation, values held by individual or groups can be drawn into a judgement so long as a clear line of reasoning is established. The problem is that there are many ways in which those with a stake in places may conceptualize environmental considerations, and give value to them in relation to social and economic concerns. Legal forms of argument, like economic ones, tend to privilege logical reasoning. There is increasing evidence that other, more intuitive, ways of reaching understanding are in common use among the different cultural communities in any society. Thus legalized argumentation, like economic reasoning, is reductionist in its treatment of the diverse concerns people have about environmental issues. This suggests that the development of discursive, communicative approaches to exploring the form and content of development plans is more likely to reflect the breadth of contemporary understanding of environmental issues than the calculative approaches being addressed by government (see Fisher 1990, Healey 1992). This would encourage plans to be prepared and implemented through broadly based processes of argumentation about actions and their impacts, which would allow discussion of both the relational concepts being used to link action and impact, and the values held about both.

The innovatory effort required to achieve the entrenchment of the new environmental agenda within the planning system is therefore substantial. It requires attention to both the substance of environmental issues and to the form and processes of planning as a *regulatory regime* for addressing collective concerns about where development should go and how local environmental resources should be managed. The overall effect, however could be to move the planning system decisively forwards to a *regulatory regime* which focuses explicitly and specifically on assessing and mitigating the adverse impacts of development projects, within a framework of precautionary limits, informed by an *argumentative* approach to planning debate which allows both technical and moral/aesthetic issues to be discussed in an open, democratic way. This would enable citizens and business to sort out ideas about appropriate thresholds and policy criteria, informed by technical knowledge and acknowledging the need to observe environmental constraints necessary to achieve targets at supra-local as well as local scales (see Healey 1992, 1993). The making of judgements is an issue for politics not techniques, and planning decisions cannot be left to an administrative/technical nexus, nor to lawyers, economists or natural scientists. The reason why the issue of democratic decision-making arises so frequently in debates on environmental issues is that the making of difficult, risky decisions needs to be widely shared among the diverse interests of a community if there is to be any chance of "entrenching" environmental criteria in stable and legitimate planning strategies which are sustained on implementation.

Thus the impact of the new environmental agenda on the planning system could be to encourage, not a reinforcement of traditional strategies and policies, but a fundamental re-thinking of its form and content, in terms of conceptions, technical methods and policy processes. As others have argued (O'Riordan 1992, Owens 1993), this could also lead to significant institutional changes, for example, to allow intersectoral coordination and stronger emphasis on regional strategy.

However, there is no inevitability about such an evolution. This paper has argued that both the meaning of environment and the approach to socio-economic-environmental relations have been contested throughout the postwar history of the planning system. The tendency has repeatedly been to contain the planning agenda to a narrow *remit* focused on land use and development and to a *discourse* dominated by forms by forms of utilitarian functionalism. The 1990s is no exception, with the pressures for opening out and for containing the environmental debate co-existing in contemporary planning debate and practice. To understand both the leverage and implications of the debates, it is essential in such times to examine both the content of the debates and the fine-grain of policy implementation. It is this latter which now needs careful attention in research on the planning system.

Acknowledgement

"Changing Meanings of 'Environment' in the British Planning System" by P. Healey and T. Shaw; copyright 1997; published in "The Transactions of the Institute of British Geographers 19(4), pp. 425-438. Reproduced by kind permission of The Institute of British Geographers.

References

Abercrombie, P. (1937) Administration of schemes prepared by joint committees. Journal of the Town Planning Institute, 23, pp 85-98.

Alison, L. (1975) Environmental Planning.

Ambrose, R. (1986) Water happened to planning Methuen, London

Ashton, J. (ed) (1992) Healthy Cities, Open University Press, Buckingham

Atkinson, A. (1990) Sowing the seeds of a green plan Town and Country Planning Vol 59 (5) pp. 1952-53

Audi Commission (1992) Building in quality: A Study of Development Control, Audit Commission, London

Banister, D. (1992) "Energy Use, Transport and settlement patterns" in Breheny, M. (ed): Sustainable development and Urban Form London, Pion pp 160-181

Beatley, T. (1989) "Environmental Ethics and Planning Theory", Journal of Planning Literature, Vol 4(1) pp 1-31

Beck, U. (1991) The Risk Society

Best, R.H. (1981) Land Use and Living Space, Methuen, London

Birles, T. Hampton, D. and Snow, D. (1992) "Planning for Energy Efficiency", The Planner Vol 87 (20) pp. 8-10

Blowers, A. (1980) The Limits of Power Oxford, Pergamon

Blowers, A. (1982), in Healey, P., McDougal, G. and Thomas, M. (ed) Planning Theory in the 1980s Pergamon, Oxford

Blowers, A. (1990) "The environment - political problem with no solution" Town and Country Planning Vol 59(5) pp. 132-133

Blowers, A. (1992) "Planning for a sustainable future: problems, principles and prospects", Town and Country Planning vol 61(5) pp. 132-135

Blowers, A. (1993) "Environmental Policy: The Quest for Sustainable Development" Urban Studies Vol 30(4/5) pp 775-796

Blowers, A. (ed), (1993) Planning for a Sustainable Environment Town and County Planning Association, London

Boehmer, Christiansen, S. (1992) "How much "science" does environmental performance review really need?" in Lykke, E. ed Environmental Performance Review: the concept and practice of environmental performance review Belhaven, London

Bosworth, T. (1993) "Local Authorities and sustainable Development", European Environment Vol 3(1) pp 13-17

Bowers, J. and Cheshire, P. (1983) Agriculture, the Countryside and Land Use, London, Methuen

Boyer, C. (1993) "Dreaming the Rational City", MIT Press, Boston

Breheny, M. (1992) "The Compact City: an introduction", Built Environment Vol 18 (4) pp 241-246

Breheny, M. (1993) "Planning the Sustainable City Region", Town and Country Planning Vol 62 pp 71-75

Breheny, M. (ed),(1992) Sustainable Development and Urban Form Pion, London

Bridnely, T. Rydin, Y. and Stoker (1989) Remaking Planning Unwin Hyman, London

Bruton, M. and Nicholson, D. (1987) Local Planning in Practice Hutchinson London

Buchanan, C. (1963) Traffic in Towns

Carson, R. (1962) The Silent Spring Houghton Mifflin, London

Commission of the European Community (CEC) 1992 Energy and the Urban Environment: Strategy for a Major Urban Centre: Newcastle upon Tyne, UK EC Brussels

Commission of the European Community (1990) Green Paper on the Urban Environment EC, Brussels

Council for the Preservation of Rural England (1993) (re South East Regional Guidance)

County Planning Officers' Society (CPSO) (1993) Planning for Sustainability CPSOS

Cullingworth, J.B. (1975) Reconstruction and Land Use Planning 1939-1947 HMSO London

Davies, H.W.E. and Rowley, A. (1986) "The relationship between development plans, development control and appeals" The Planner Vol 72 (10) pp 11-15

Davoudi, S. (1993) Development Plans and sustainable development paper to AESOP Congress, Lodz, Poland, July

Delafons, J. 1991 Planning in the USA - paying for development The Planner Vol 77(20) pp. 8-9

Department of the Environment (1974) Circular 98/74 Structure Plans HMSO London

Department of the Environment (1975) Review of the Development Control System (Dobry Report) HMSO, London

Department of the Environment (1977) Circular 55/77 Memorandum on Structure and Local Plans HMSO, London

Department of the Environment (1979) Circular 4/79 Memorandum on Structure and Local Plans HMSO, London

Department of the Environment (1984a) Circular 14/84

Department of the Environment (1984b) Circular 15/84

Department of the Environment (1985) Circular 14.85 Development and Environment HMSO, London

Department of the Environment (1986) The Future of Development Plans: a Discussion paper DoE, London

Department of the Environment (1988) Planning Policy Guidance N° 12 Local Plans HMSO London

Department of the Environment (1990) This Common Inheritance. HMSO, London

Department of the Environment (1992) Planning Policy Guidance 12: Development Plans and Regional Guidance HMSO, London

Department of the Environment (1993) The UK strategy for sustainable Development Consultation Paper, DoE, London

Department of the Environment/ Department of Transport (1993) Reducing Transport Emissions through Planning HMSO, London

Despotakis, V., Giaoutzi, M. and Nijkamp, P. (1992) "Spatial Depiction of Local Sustainable Development" in Breheny, M. (ed): Sustainable Development and Urban Form Pion, London

Druker, H., Dunleavy, P., Gamble, A., and Peele, G., eds. (1985) Development in British Politics, Macmillan, London

Dryzek, J.(1990) Discursive Democracy, Cambridge University Press, Cambridge

Elkins, P., ed.(1986) The living Economy: a New Economics in the Making, Routledge and Regan Paul, London

Elkin, T. and McLaren, D. (1991) Reviving the City, Friends of the Earth, London

Elson, M. (1986) Green Belts, Methuen, London

Fisher, F. (1990) Technocracy and the Politics of Expertise, Sage, London

Flyvberg, B. (1992) Rationality and Power, Academic Press, Cophenagen

Friedman, J. (1992) Empowerment: the Politics of the Alternative Development, Blackwell, Oxford

Friends of the Earth (1989) Environmental Charter for Local Government, FoE, London

Grant, M. (1992) Planning law and the British land use planning system, Town Planning Review, 63, 3-12

Goodchild, B (1990) Planning and the modern/postmodern debate, Town Planning Review, 61/2, 119-137

Hajer, M. (1992) The politics of environmental performance review: choices in design in Likke, E. (ed.) Environmental Performance Review, Bellhaven, London

Hall, P., ed. (1973) The Containment of Urban England, Allen and Unvin, London

Harvey D. (1989) The Condition of Postmodernity, Oxford University Press.

Healey, P. (1993) "Discourses of Integration: Making Frameworks, processes and Contents for democratic Urban Planning", Paper presented in Challenges in Urban Management conference, University of Newcastle, 25-27 March

Healey, P. et al. (1988) Land Use Planning and the Mediation of Urban Change. Cambridge University Press, Cambridge

Healey, P. (1992) The reorganisation of state and market in planning, Urban Studies 29, 411-434

Healey, P. and Shaw, T. (1993) The treatment of environment by planners: evolving concepts and policies in development plans, Working Paper, Department of Town and County Planning, University of Newcastle upon Tyne

Jacobs, M.and Stott, M. (1993) "Sustainable Development and the Local Economy", Local Economy 7/3 pp. 261-272

Lichfield, N. (1989) From Planning gain to community benefit, Journal of Planning Environment Law, 68-82

Lykke, E., ed. (1992) Environmental Performance Review, Bellhaven, London

McLaren, D. (1990) The scope for planning, Town and County Planning, 59, 148-50

Nijkamp, P. (1990) Sustainability of Urban Systems, Newcastle upon Tyne.

Nijkamp, P. et al.(1990) "Project appraisal and Sustainability". Seminar Act, Wagenigen.

O'Riordan, T. (1990) Environmentalism, Pion, London.

Newby, H. (1990) "Ecology, Amenity, and Society", Town Planning Review 61/1, pp.3-13

OECD (1993) Ecological City. OECD, Paris.

Owens, S. (1992) "Energy, Environmental Sustainability and Land-use Planning" in M. Breheny, M. (eds.) Sustainable Development and Urban Form, Pion, London, pp.79-105

Pearce, B. (1992) The effectiveness of the british land use planning system, Town Planning Review, 61, 3-13

Pearce, D., Markandya, A. and Barbier, E. (1989) Blueprint for a Green Economy, Earthscan, London

Rydin, Y. (1993) The British Planning System. Macmillan, London

Thornley, A. (1991) Urban Planning under Thatcherism, Routledge, London

WCED (1987) Report of the UN World Commission on Environment and Development: Our Common Future, Oxford University Press

Whatmore, S. and Boucher, S. (1993) "Bargaining with nature: the discourse and practice of 'environmental planning gain' ", Transactions of the Institute of British Geographers 18/2, pp. 166-178

Williams, J.(1993) Environmental Opportunities: Building Advantage Out of Uncertainty. Centre for Exploitation of Science and Technology, London.

Wood, C (1989) Planning Pollution Control, Heinemann, London

Notes

[1] see This Common Inheritance, Secretaries of State 1990.

[2] (Add: Words of 1991 Act re plans)

[3] The new environmental ideologies challenge in particular modernist preoccupations with material growth and economists' concentration on evaluating the financial costs and benefits of policies and projects. See Hajer 1992, Dryzek 1990, Flyvberg 1992, Harvey 1989.

[4] see the World Conservation Strategy 1980 produced by the International Union for the Conservation of Nature and National Resources, Gland, Switzerland.

[5] In these ideas which are already vigorously developed in policy debate in Northern Europe (see Desptakis et al 1992, TRP).

[6] The term discourse is here used to refer to a system of meanings and metaphors.

[7] see The Planner and Town and Country Planning from 1990 onwards, the work of Owens, Marshall and Rydin, COPs 1993, Blowers ed 1993, DoE/DTp 1993.

[8] This common Inheritance (DoE 1990) was the Government's first exercise in greening its overall policy agenda.

[9] See statements by Minister Howard in 1993 (Planning 1014) and DoE/DTp 1993)

[10] This argument was used by the British government against the introduction of Environmental Impact Assessment legislation by the EC.

[11] see Blowers 1980

[12] Rydin (1993) develop an interesting analysis of contemporary environmental discourse in the planning system in term of 'mundane' and 'sublimne' elements.

[13] Their views appear to reflect a Baconian conception of the relation between people and nature (Harvey 1989)

[14] See the various "Countryside in 1970" activities during the 1960s, culmination in the ceation of the Countryside Commission.

[15] For example, the Sandford Committee on natural parks, the Verney report on aggregates and the Porchester report on Exmoor

[16] These centred on the realization of limits to environmental capacities, and the long-term consequences of short-term environmental neglect. The Aberfan tradedy, and publication of Rachel Carson's Silen Spring (Carson 1962) and the Limits of Growth (ref) study helped to frame these concerns. Note also the impact of the OPEC oil crisis on energy conservation.

[17] The Strategic Plan for the South East 1970 predicted that sufficient land had already been allocated in the late 1960s to accommodate development until 1980.

[18] Despite the innovative activity of some of its members, the RTPI was very slow to take up the issue of environmental sustainability. The Town and County Planning Association, in contrast, vigorously promoted debate, culminating in a substantial policy report on Planning for a sustainable Environment published in 1993 (Blowers 1993).

[19] The contrast between DoE guidance on development plans in 1988 and 1992 illustrates this clearly (DoE 1988, 1992).

[20] Northern European planning debate and practice were significantly ahead of British discussion, see Nijkamp et al 1990).

[21] The discussion of development impacts was in any case been given prominence as a way forward in the long-standing debate on "planning gain", see Delafons 1991, Healey et al 1993).

[22] "Conservation and development should not be seen as necessarily in conflict. Policies for land use must weigh and reconcile priorities in the public interest." (para 6.4) "Environmental concerns weigh increasingly in the balance of planning considerations" (Para 6.7).

[23] see Planning 1014, 1993

[24] Work in Sweden and the Netherlands provides helpful ideas in this regard (see Nijkamp 1990, Nijkamp et al 1990)

Patsy Healey, Tim Shaw
Centre for Research in European Urban Environments
University of Newcastle upon Tyne
NEI 7RU
United Kingdom

14 LANDSCAPE EVALUATION AND PLANNING IN THE VENETO REGION

Giorgio Franceschetti
Tiziano Tempesta

14.1 Introduction

The Veneto Region has for a long time focused its actions on the protection of the landscape.

To reach this goal, the Region has started many interventions, of which two are of maximum importance: the Regional Master Plan (*Piano territoriale regionale di coordinamento* or P.T.R.C.) and the setting up of Regional Parks (Regione Veneto, 1991).

Through the Regional Master Plan, the Veneto Region has defined:

- the most important strategies for landscape policy;
- the areas where parks or natural reserves will be realized or that must be the object of detailed landscape planning;
- the elements of landscape that will have to be protected by the provincial and municipal plans.

After adopting such measures, the Region has enacted special laws that protect some areas. The first of them include the regional Euganean Hills park (*Colli Euganei*) that has the main purpose of protecting the rural landscape of a hilly area that is situated in the centre of the Veneto Plain.

The Regional Master Plan is certainly an important instrument of landscape protection and improvement but some aspects of it are perplexing.

Firstly, the Regional Authority has not analysed whether or not there is a demand for landscape and, if there is, what qualitative features of the territory affect it. Secondly, an analysis of the possible economical effects of the parks and reserves has been left completely undone.

The present work gives the results of two items of research recently carried out in the Veneto region[1]. In the first, we compared the goals of the Regional Master Plan for landscape protection with what individuals prefer. To reveal their preferences about landscapes, we made a survey in the province of Treviso that encompasses the most important landscape of the Veneto region. The second study assessed the effects of the

D. Borri et al. (eds.), *Evaluating Theory-Practice and Urban-Rural Interplay in Planning*, 193–207.

Colli Euganei park on the land values, with the aim of verifying whether the landscape protection instruments can affect real-estate values.

14.2 Individual preferences and public aims of protecting the landscape

The goals of the P.T.R.C. on the landscape protection in the province of Treviso.

To protect the landscape, the Regional Master Plan defined the landscapes that will have to be recognized and protected by subordinate (provincial and municipal) plans. With reference to the province of Treviso, the plan indicated the following landscapes:
- the landscape of the *cavino* or paduan layouts (which are ancients layouts in the Veneto lowlands);
- the vinicultural landscape of the low hills;
- the water meadows and the permanent grasslands;
- the arable landscape with vine plantings (*piantata di viti*).

Secondly, the Regional Master Plan defined where parks and nature reserves should be established, including the following of landscape significance:
- the Sile river springs park;
- the *Palù del Quartier del Piave* park.

Finally, the Regional Master Plan indicates the area of the Montello hill as a location for a territorial recreational park.

A survey of community preferences was made on each of these landscapes (Tempesta, 1993), in which 27 different pictures of the province of Treviso were submitted to the judgement of 107 people. The pictures were chosen with the following thoughts in mind:
- to submit to people's judgement the landscape elements that will be protected;
- to represent all the most important landscapes of the province;
- to show some similar landscape but with one or more different features;
- to submit only ordinary landscapes to people's judgement.

The pictures were taken in the following landscape ambits (Marchetti, Milani, Santantonio, Tempesta, 1991):
- the low southeasterly plain with a prevailing diffusion of large capitalistic farms;
- the low southwesterly plain with small farms;
- high plains where sometimes there can still be found the signs of Roman land partitions;
- the low easterly plains with a strong diffusion of viniculture;
- the Asolo hills;
- the Montello hill;
- the hills between Valdobbiadene and Vittorio Veneto;
- the high hills and prealpine mountains;
- the closed fields of the Piave District.

The evaluation methods

In this study we considered two different ways of evaluating landscapes:
- scores reflecting aesthetic taste;
- a monetary evaluation of recreational behaviour, centred on the willingness to travel.

The first has often been used and can give very useful indications of the factors affecting the aesthetic appeal of each landscape.

To get a monetary estimate of the recreational value of given landscapes we used the willingness-to-travel approach and asked our subjects to say the maximum travelling cost they would be willing to pay in order to visit all the 27 landscapes, without reducing the number of the visits. After that, by subtracting the real travelling cost from the hypothetical we calculated the consumer surplus, which we took to equal subjects' willingness to travel, clear of travelling expenses.

Estimating the aesthetic-recreational value of the landscapes of the province of Treviso

Our analysis of the gathered data shows a relationship between the development of recreational activities in rural areas and the landscape features. Although there are some important exceptions, we observed that the most frequented landscapes are also the ones mostly appreciated aesthetically.

The hilly and the plains landscapes differ markedly in aesthetic appeal. As far as the plains are concerned, we must underline the high aesthetic-visual value of the *Palù del Quartier del Piave* landscape, an area of small fields enclosed by thick hedges. This shows that even flat landscapes are appreciated when they contain considerable arboreous vegetation. It is interesting, in this connection, to compare the evaluations of this landscape with that of the Montello hill area, where present features are due to land reclamation at the beginning of the century. The plains and their historical landscapes are appreciated as much as the hilly ones if the latter have modern landscapes without too many diversifying elements.

On the other hand, plains score higher when they include watercourses rich in arboreous vegetation or bushes.

As far as the hills and the mountains are concerned the landscape of the Asolo area is the most appreciated. Because of the particular morphology of this landscape, many of its traditional elements have been maintained. The low productivity of its soils has in fact prevented the spread of intensive viniculture, and its agricultural sector is still based on mixed systems of viniculture and zootechnical farming.

Contrary to expectations, the most typically Trevisan mountainous landscapes attracted less interest perhaps because their mountains aren't as attractive as the Dolomites to their north.

Also our estimate of consumer surplus per trips, made with the willingness-to-travel method, yielded similar results. The monetary-recreational value gives results very close to those of aesthetic judgement, although the data are much more articulated. In the plains the surpluses per trip vary from 0 to about 4,000 lire for the river areas; many areas wholly lack any recreational value, especially those that are remote far from major built-up areas or that lack environmentally interesting elements. On the contrary, all the values for the hilly areas exceed 3,000 lire per trip with peaks of 16,000 lire for the Asolo hills and 23,000 lire for the alpine-hut landscape. Furthermore the mountains and high hills usually have a much geater monetary value than the subalpine hilly zone.

Factors affecting the landscape value.

The values of the aesthetic-visual index and of the consumer surplus per trip sometimes differ considerably even for landscapes with similar features. So it seemed interesting to identify the landscape characteristics that could cause such a variations; we used a stepwise regression.

We identified ten factors that can completely describe each landscape: 1) morphology, 2) visual width, 3) crops, 4) vines, 5) grassland, 6) woods, 7) qualifying elements (like rows of vines, historically interesting rural buildings, etc.), 8) agricultural layout of historical interest, 9) rural buildings, 10) hedges.

Besides these, other variables refer to the quality of a given picture: its brightness, the season it shows and its contrast with the picture shown immediately before it to our subject. In this way, we could verify whether and how much the photographic techniques and the order of presentation could have affected values.

Once we had defined the means of predicting landscape values, we made our stepwise regression analysis and found some possible explanatory models for our two indexes, which are shown in Table 14. 1. The first element revealed by the models is that a landscape's morphology is basic to determining both its aesthetic and recreational-monetary values. Other factors being the equal, a hilly landscape has an aesthetic score almost 40% higher than that of a plain landscape (model 1). We must observe that landscapes are appreciated for some features of their agricultural use, such as grassland and hedges. In fact these two elements considered together assume almost the same importance for aesthetic value as morphology.

The consumer surplus per trip and morphology correlate with the presence of grassland, of rural buildings (model 2) and of the depicted season (model 3). Even in this case besides the strictly physical factors, such as hills, even typical elements of human intervention seem to influence recreational value. In the models, the presence of rural buildings reveals that people prefer areas where there "built-landscape" signs to "wilderness" areas. This preference for a "nice" to a natural landscape follows a cultural line of a humanistic-renaissance kind of behaviour that is very strong in this country.

P.T.R.C. aims and individuals' preferences: a few conclusions.

As developed, our analysis showed the main factors that determine the aesthetic or recreational value of a landscape are elements of great visual importance and include its morphology and the presence of grasslands, woods, hedges and rural buildings. But we found that other historical-cultural and environmental elements were not important. So none of our statistical models featured the presence of elements typifying the historical landscape (e.g., rows of vines, historical agrarian layouts, etc.).

Our analysis showed that the aims of the Regional Master Plan and people's preferences differed very greatly. First of all, our interview subjects had no aesthetic or recreational interest in some of the historical elements of the landscape that the Veneto Region wants to protect; two particular elements of the historical Veneto landscape, its rows of vines and *cavino* layout, were not particularly prized.

As to hilly landscapes, the Regional Master Plan does not mentioned the one most preferred, for the Asolo hills lack any landscape element that the Plan is concerned to protect.

Table 14.1 Interpretative models of the values of the aesthetic-visual index (IEV) and the consumer's surplus-per-trip index (SURP).

Model 1

IEV = 1.30 MORPHOLOGY + 1.25 GRASS+ 0.71 HEDGES + 3.40
(7.274) (4.168) (2.468) (13.462)

$r2 = 0.85$

Model 2

SURP = 4219 MORPHOLOGY + 5518 BUILDINGS + 3754 GRASS - 1650
(5.357) (3.870) (2.574) (-1.421)

$r2 = 0.75$

Model 3

SURP = 4897 MORPHOLOGY + 5350 BUILDINGS + 3236 SEASON - 2208
(6,672) (3.643) (3236) (-1.529)

$r2 = 0.74$

All coefficients are significant at 5%

Paradoxically, great attention has been given to a hilly landscape of slight popular interest, the Montello area, where the landscape has deteriorated because suburbs have sprawled out into the countryside.
Thus in many respects, the Regional Master Plan seems wrongly directed and only partially in touch with the community's preferences.

14.3 The effects of landscape protection on real-estate values: the case of the Colli Euganei Regional Park

Purposes

The foundation of a park is general followed by the introduction of some restrictions on land-resource uses, which may reduce the benefits derived from using the resources. So founding a park may strongly affect real-estate values.
A study of the changes in the land market can interestingly indicate the effect of a park on real estate and, therefore, on the income of the people working within the park area (especially farmers).

In order to verify such problems, we examined the evolution of the land market in an area of the Colli Euganei park and in other areas outside it. This allowed us to compare the land market before and after the park was set up and to formulate hypotheses on the emerging elements of diversification.

The investigation methodology

The sources of information about the Italian land market are meagre. In the Veneto Region, no systematic service gathers the main factors of the market (prices, details of the real estate that changes hands, contractors). Since we intended to go deeply into what can be known about the *Colli Euganei* land market, we investigated the relevant contracts of sale sent each month from the Registrar's Office to town councils, where the INVIM tax is calculated. This produced detailed information on, for example, the kinds of real estate sold, some of their physical characteristics, the categories of the subjects involved, etc.. However, most data on the real-estate prices were unreliable (Fratepiero, 1990).

In point of time, we investigated three periods significant for the introduction of restrictions affecting the park:

1982-1985: no restrictive laws existed;

1986-1988: in 1986, the Regional Master Plan established the temporary borders of the park and some provisional land-use restrictions;

1989-1992: a specific regional law (38/1989) established the park and many restrictions came into force.

To isolate the effect of the founding of the park on land-market trends, we investigated transactions in land both inside and outside the protected area of the park. Within the park, we considered an agricultural and forest area of 3484 hectares (24% of the whole park) and an area of 2315 hectares outside it. The contracts of sale concerned a total area of 1275 hectares.

The results of the investigation

Our investigation identified the main elements that diversified the trend of the land market inside and outside the park; they are as follows:

- once the Colli Euganei area had been designated as a park, the market in this hilly land became more mobile (figure 14.1), especially for land with buildings;
- a second element of strong diversification of the hilly land market with respect to the plain is the importance given to lands equipped with rural buildings. Of sales of land within the park, an average of 45% were of cultivated land with buildings but in the plains ch sales never exceeded 30% of the total;
- during the 1980s, much land was sold by non-residents to residents, a tendency that favoured farmers and other professionals. This process was diversified inside and outside the Park. Outside the park, such sales decreased during the observed period, while outside it they increased. Once the area was designated as a park, sales of land by non-residents to residents was emphatic;
- land-market values were fundamentally cyclical, although they differed interestingly inside and outside the Park (as figures 14.2 and 14.3 show).

Figure 14.1: Sales of land as percentages of the total cultivated areas in the (hilly) park and on the plains.

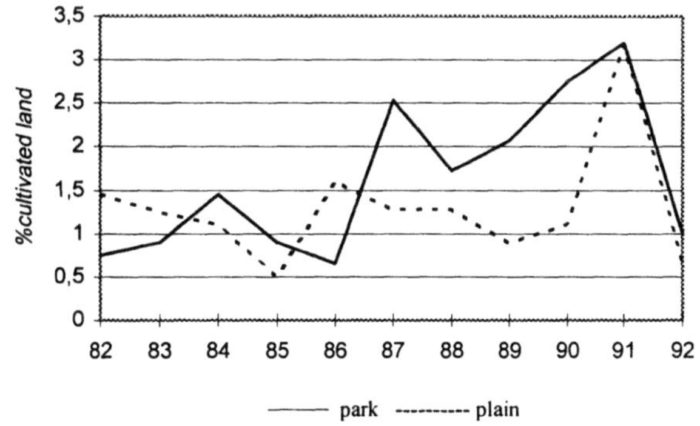

Figure 14. 2: Average prices of the arable land in the park and in the plains between 1982 and 1992.

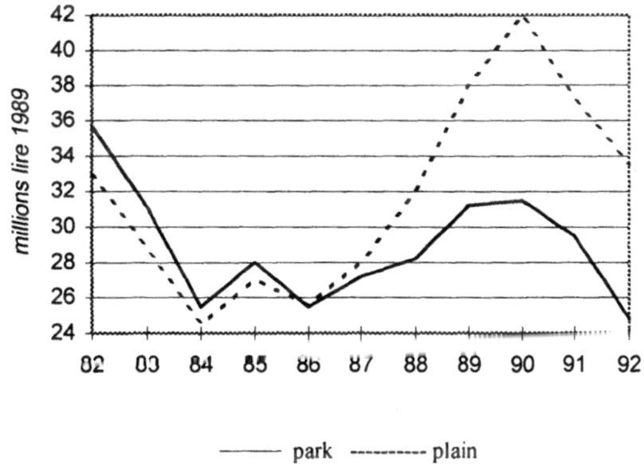

Figure 14.3: Average prices of the vineyards in the park and in the plains between 1982 and 1992.

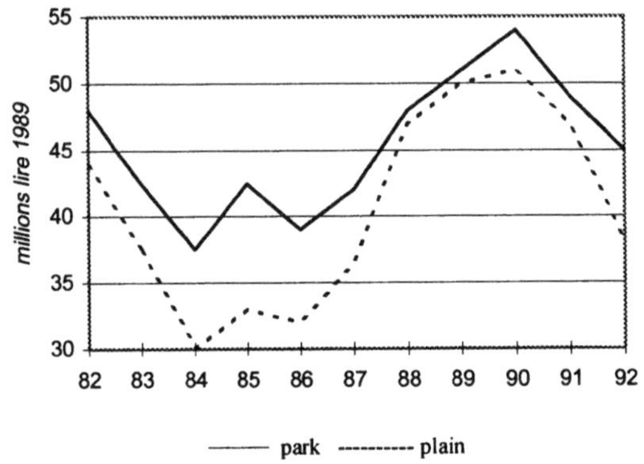

Both the deflated values of the arable land and of the vineyards tended to diminish up to 1986, to increase until 1990 and to diminish during the following two years (figure 14.2 and figure 14.3). The arable land and the vineyards inside and outside the park showed the following average values:

year	arable land		vineyard	
	park	plain	park	plain
1982	35.7	44.2	47.9	52.6
1986	25.4	33.5	39.1	39.4
1990	31.5	47.8	53.6	53.6
1992	24.6	35.8	45.0	40.1

(1989 millions of lire)

The values of farm land with rural buildings were different. The first point to emerge is that their values have always been much higher in the hills than on the plains. The average values of a standard farmland of 1500 m^2 with a 600 m^3 building to be refurbished located inside and outside the park were:

	Park (A)	Plain (B)	A-B
1982-85	80	34	46
1986-88	102	39	63
1989-92	115	44	71

(1989 millions of lire)

Not only the value of the buildings is higher inside the park, but it has also increased very much over time. We note the great variation found in the area of the park before and after it was designated as a park and the further increase in 1989-1992. No such increase affected land on the plains.

A final consideration is variations in value of different kinds of land in the periods examined:

	arable land value		vineyard value		land value of 1500 m² with build.	
	park	plain	park	plain	park	plain
% variation 1982-85/1986-88	-10.6	-2.2	-2.2	-3.3	+27.0	+14.7
% variation 1986-88/1989-92	+7.4	+34.2	+19.8	+29.1	+12.4	+12.8

From this, we deduce that, while land without buildings inside the park fell in price more than such land outside it, rural buildings appreciated in value, which demonstrates an asymmetrical effect on land-market values of the foundation of the park.

Foundation of the Colli Euganei Park and the market in land: some interpretative suppositions

In order to better understand the trends shown above, we have first to consider that interest in land in rural areas falls into three main categories (Grillenzoni, 1981):
- for housing;
- for agriculture;
- for financial speculation.

While the financial category is mainly responsible for the cyclical features of the market, demands for housing and cultivation may have caused the most important differences in the trends inside and outside the park.

As to sales of land without buildings, Figure 14. 4 shows how, for land in the park area, market changes altered between the first half of the 1980s and 1987-91 on account of an expansion of both demand and supply. The market-equilibrium points in fact show a shift to the right.

During the same period (1982-92) in the plains (Figure 14. 5), demand expanded but supply remained steady, and we can suppose that the founding of the park led certain categories of owners to sell their land mainly because of the restrictions on building activity. In fact, mainly non-residents, being people mainly interested only in building activity, determined the rise in land-market mobility during the late 1980s. The rise in demand during the period in question shows, on the other hand, that the agricultural operators were not pessimistic as to the effects of the park on agricultural incomes.

As to the plains, the farmers bought more and more of the land offered for sale.

As to sales of lands with buildings (Figure 14. 6), it is possible to notice even here that events in the park differed from those on the nearby plains. Even though in the market rose for land in both areas, it is possible to see that the market rise in the park is due mainly to the increase in demand, for supply remained unchanged. In the *colli Euganei*, the rise in value of land with rural buildings related to a loss of building possibilities on the remaining rural territory. Part of the demand for land for residential purposes that had led to medium-large pieces of land changing hands has been directed by building restrictions towards land with rural buildings.

In conclusion, we can emphasize that founding the park did not penalize the residents, who were mainly farmers, at least during the first period. Moreover, rural buildings have increased greatly in value, which favours the resident population, even if a good number of the rural buildings in the hills belong to people who live on the plains.

To better investigate the effects of variations in land prices, we estimated their values during the 1980s (Table 14. 2).

The value of real estate in the park area rose from 710 billion lire in 1982 to 943 billion lire in 1991 at 1989 constant prices (+ 32%). Buildings constitute the most important part of the whole, and their share of the total rose from 56% in the 1982-1984 period to 63% in 1989-1992. Besides buildings, the values of land with vines and wooded land rose, and those of arable lands and pastures fell during the period.

Final remarks

The founding of the park has not reduced rural land values. Although we cannot assert that it is the sole cause of all the estate changes, it is possible that, in some way, it helped to increase real estate values of inherited buildings and so compensated for losses in value from diminished building possibilities on agricultural land. On the other hand, our analysis gave prominence to how the recreational use of the land, and more specialized, professional forms of agriculture, such as viniculture, have increased in the park thanks mainly to the intense touristic interest.

Figure 14.4: Average values of land without buildings and land-market mobility in the Colli Euganei Park between 1982 and 1992.

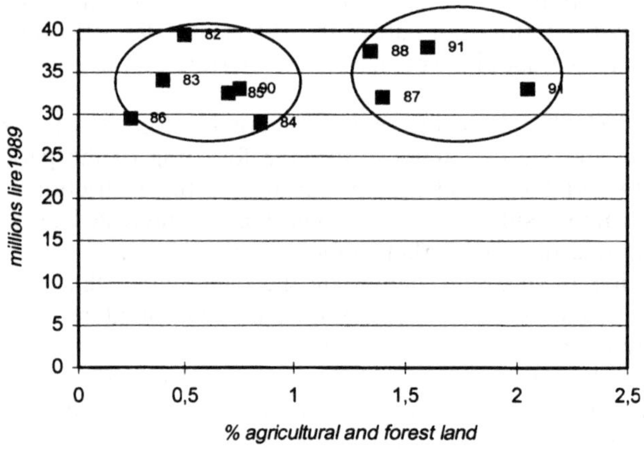

Figure 14. 5: Average values of land without buildings and land-market mobility in the plains between 1982 and 1992

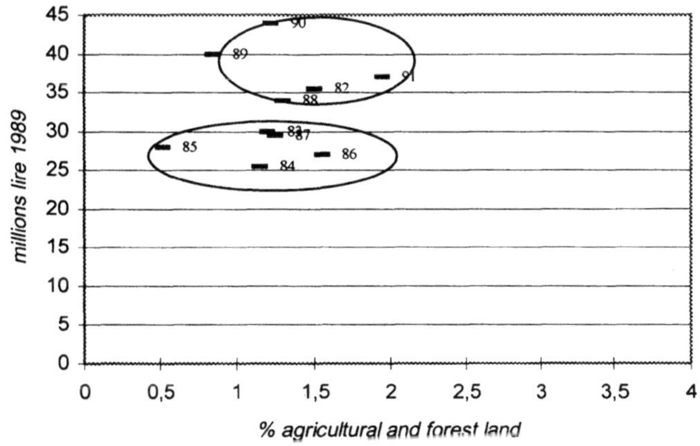

Figure 14.6: Lands with buildings sold and average land values in the plains and in the Colli Euganei Park between 1982 and 1992.

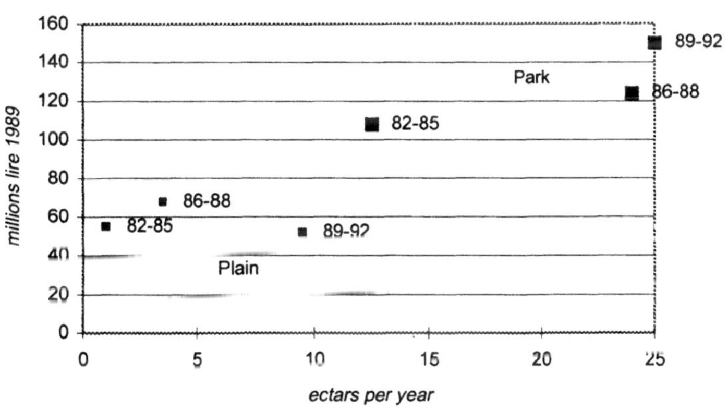

Table 14.2 Real-estate values in the Colli Euganei park in 1982, 1986 and 1991 (milion lire 1989).

1982

	surf. ha	price million	value million	%
arable land	3252	30.4	98871	13.9
vineyard	3498	42.5	148651	20.9
grassland	651	21.4	13932	2.0
woodland	5322	7.4	39384	5.5
uncultivated land	1101	8.7	9576	1.3
total surface	13824		310413	43.7
buildings (number)	5000	80.0	400000	56.3
total value			710413	100.0

1986

	surf. ha	price million	value million	%
arable land	3082	27.2	83841	10.7
vineyard	3454	31.9	110191	14.0
grassland	610	20.3	12377	1.6
woodland	5362	8.8	47182	6.0
uncultivated land	1108	8.2	9089	1.2
total surface	13616		262679	33.5
buildings (number)	5120	102.0	522240	66.5
total value			784919	100.0

1991

	surf. ha	price million	value million	%
arable land	2870	29.2	83804	8.9
vineyard	3400	49.7	168980	17.9
grassland	558	22.3	12460	1.3
woodland	5411	12.8	69261	7.3
uncultivated land	1118	10.1	11292	1.2
total surface	13357		345797	36.6
buildings (number)	5200	115.0	598000	63.4
total value			943797	100.0

After all, we can affirm that, in the *colli Euganei* area, many current land-use transformations tend to give more importance to environmental-landscape aspects that favour a revaluation of real estate and, in many cases, increase the real estate values of residents and - mainly - of farmers.

14.4 Conclusion

The two case studies show how planning policies for landscape protection are sometimes not well aimed and how they can determine very strong redistributive effects.

The difficulty of identifying correctly the social preference functions referring to the use of public goods is one of the main limitations of the interventions made in the Veneto region to protect the landscape. Since the mid 1980s, the region has adopted many measures to control changes in land use and to administer the private use of environmental resources.

These interventions had the general aim of defining plans that would have precise goals and articulations only in a second stage that followed political mediation; the hypothesis was that plan administrators could correctly perceive social preferences.

Such a hypothesis is not correct because politicians have their own ideas of "utility functions" as means of raising their political roles and maintaining their "political lives". Maximizing a politician's utility function coincides only partially with the maximum welfare of the community: in other words, a search for electoral consent does not necessarily maximize the community's well-being.

Even in a democratic society, its political class often meets great difficulties in giving the right weight to the different requests that come from the rest of society.

The interests of strongly motivated minorities can assume a greater importance in the eyes of the politician than those of the majority of society. Typically, when a few people would suffer greatly from a given act while others in the community would be slightly favoured individually, but greatly favoured as a whole, what is done is rarely to maximize the community's welfare, for it is much easier for the politician to accord greater importance to the needs of the strongly motivated minority.

Although this choice may be wrong for the welfare of the community, the politician will see it as best because it favours "his individual welfare"! Slight harm to most people will not lead to any political changes while a few greatly harmed people can adopt countermeasures that could be negative for the politician who has made the harmful decisions.

Moreover, a second element tends to divert political choices away from maximizing the welfare of the community: the non-measurement of the real redistributive effects of the choices made. In the Veneto region, the adoption of most of the laws and plans regarding the landscape and the environment were not followed by any study of putative changes in income of different socio-economic categories. Without such studies, the relevant politicians are often led to consider that the real redistributive effect is the one felt by the individuals directly affected by the plan. As a result, the effect is a tendency to over-value the interests of those whom the plan seems to harm, and under-value the interests of those benefitted by the plan. According to studies conducted by Kahneman and Tversky (Mitchell, Carson, 1989, p. 35), people are usually most adverse to reducing their goods and resources. According to their "prospect theory", the demand function would have less sensible results when the availability of a benefit is reduced rather than raised. Such a gap is greater for all the utilizable environmental resources whether used for the production of saleable goods or to satisfy primary or cultural needs (e.g. recreative activities). It is nevertheless important to remember that, while the benefits that go to those who use the resources for productive purposes are easily measurable, the estimates of those regarding the recreative and cultural use is much more difficult to quantify.

For such reasons, political actions and choices about the protection of the environmental and cultural goods will tend to favour the status quo and to support the present use of the resources (generally of a productive kind) to the detriment of the valorization-transformation for cultural and recreative purposes or of the interests of future generations.

Such considerations allow us to state, furthermore, that besides the well-known failure of the market in allocating resources, we may speak of a "failure of the public operator" in attempting to remedy the poor operation of the market. Such a failure, like that of the market, may be considered intrinsic to the mechanisms that are fundamental to public choices made in modern democratic societies, at least to how they have been done until now. Therefore the allocation processes created by market mechanisms and the political decision-making processes share the need to be corrected especially in some categories of environmental resources.

After all this, it appears essential to think over how laws on the environment and planning the use of resources have been conceived and carried out. Although the conditions for the optimal use of resources will never be reached, it is clear that we could get more efficiency by a careful use of economical analysis. Particularly for environmental goods, a great effort should be made to qualify and quantify possible alternative use demands. Techniques that aim to analyse the demand for environmental and landscape resources have been tested for more than twenty years. Even if they have important, known limits, they can nevertheless offer an important analytic and cognitive support for decision makers.

References

Fratepiero G. (1990), Accertamento di Valore e Trasparenza dei Mercati, Genio Rurale 7/8

Grillenzoni M. (1981), Il Valore della Terra, Edizioni Agricole, Bologna

Marchetti, G. Milani, G. Santantonio, O. Tempesta, T. (1991) Il Paesaggio Agrario della Provincia di Treviso, Treviso

Mitchell R.C., Carson R.T.(1989), Using Surveys to Value Public Goods. The Contingent Valuation Method, Resources for the Future, Washington D.C.

Regione Veneto (1991), Piano Territoriale Regionale di Coordinamento, Venezia.

Tempesta T. (1993), “La valutazione del paesaggio rurale tramite indici estetico-visivi e monetari”, Genio Rurale 2

Notes:

[1] The present work was planned by the authors jointly: G. Franceschetti wrotes sections 1 and 4, while T. Tempesta wrote sections 2 and 3.

Giorgio Franceschetti
Tiziano Tempesta
Dipartimento Territorio e Sistemi Agro-Forestali
Via Grandenigo 6
35131 Padova
Italy

15 EVALUATING FUNCTIONS IN URBAN-RURAL AREAS

Giovanna De Fano
Giovanni Grittani

15.1 Introduction

There is a general consensus among the scientific community and the general public that rural areas on the periphery of towns play a strategic role in regional planning.
Terms like *sustainable development*, *better quality of life*, *etc.*, which have become widely used, show a growing concern for the preservation of environmental resources and for urgent decisions about regional planning and management.
Everybody is aware of the negative impact on urban peripheries of conflicting land use where economic and social structures have been jeopardized and the other environmental functions lost.
When one thinks of a well-known rural landscape, it is easy to focus on why it should be preserved. First of all it simply is the land where agricultural activities are carried out, activities that are obviously usually connected with some specific social, historical and cultural elements, tangible evidence of models of life and thus significant symbols of the historical heritage of future generations. If such a context is located in an area that is easily accessible to the public, as is any rural periphery of a town, its value is even greater. Consequently, it is important to adopt a planning strategy geared to preserve the rural and environmental features of such peripheral farmland.
The continuous and chaotic spread of urban centres has largely affected farmland, which has suffered an indiscriminate loss of acreage. In Italy, planners have not taken effective account of the productive and environmental features of peripheral agricultural areas (Brugnoli, 1990; Toccolini, 1990; Zappavigna and Tagliavini, 1990).

15.2 The functions of peripheral farming

In most cases the production structure of peripheral farming is weak. In areas far from the city, productivity is the main reason for continuing agricultural activity but in peripheral areas the farmers' usually merely wait for building plans to upgrade their lands, which is the "impermanence syndrome" (Lockeretz, 1989). A symptom is that

D. Borri et al. (eds.), Evaluating Theory-Practice and Urban-Rural Interplay in Planning, 209–215.

farmers tend to reduce investments in farming, which makes the future of agriculture even more uncertain in these areas. This is the last step before land becomes idle. Under the impact of generational changes, too, the future of agriculture seems even more uncertain.

Given the weakness of agrarian production processes, regional and sectorial planning should focus on action to limit and compensate for the causes of impermanence in agriculture, in order not to break the connections between towns and the countryside and to maintain an acceptable quality of urban life and in near-by rural areas. Farmers near towns are generally thought of only as producer of foodstuffs, only rarely as indirectly stemming environmental degradation.

There are various levels of analysis of urban-peripheral land. The points of view of people not directly engaged in farming should be considered. Some members of the public are concerned with either preserving rural spaces or converting farmland to non-agricultural uses.

Members of the general public tend to consider peripheral farming as generating mainly recreational and aesthetic externalities. They imagine the countryside as offering what the city does not: clean air, beautiful landscape, a quiet life, genuine products. They tend to emphasize the positive aspects of the countryside, as opposed to the negative aspects of the city. In general they tend to support the preservation of peripheral farming.

Some people, including real estate agents, building contractors and landowners, consider such land as the basis for urban development and have a keen interest in a rapid spread of the urban area and obviously in lucrative rents to be got through urban development of the land. In this conflict, politicians should arbitrate, being helped by the planner's technical support. Obviously their role should be neutral because of the different interests involved.

However, observing what has happened in our country, it should be noted that politicians have not been at all neutral in making planning choices. Their behaviour has been favoured by the lack of autonomy of urban planners and by an urban view of urban development. The reasons are still unclear: a cause may be a lack of technical and cultural experience or money-making tactics.

Urban plans have wasted rural areas by overestimating spatial needs. Inflated plans create "empty" spaces that are unlikely to become part of the "city": they end up fostering the "impermanence syndrome".

Politicians have not actually been able to grasp the productive and environmental value of rural areas near towns.

Agricultural activities practised on urban peripheries are a sort of "connective tissue" between towns and the countryside. In this particular location, no definite, precise line can be drawn between town and country. Hence, when planners and politicians readily assert that they can clearly separate the two, their position may defined as overtly ambitious.

14.3 The evaluation approach

Revising approaches to planning is definitely the most important issue in developing and properly using peripheral rural areas, and a more coherent understanding of farming in them is the first essential step. New town-planning "rules" should be sought which deem farming in these areas to comprise production, landscape and recreation. As long as farming space is considered residual, towns will be served very poorly: green areas (and many other "signs") will be lost and not preserved and a negative impact will affect productive and social structures in these peripheral areas.

A definite analysis of farming near towns would require rural areas to be carefully classified into their various functions. It would be advisable to map the various types of urban and rural land use to assess their productive, recreational and landscape values.

Firstly, a set of indicators should be established to express the various functions attributed to rural areas. Secondly, the different aspects of evaluation concerning interactions between planning action and the agricultural setting should be investigated by means of a descriptive analysis as defined by the relevant indicator.

Because of the manifold levels of value attributed to peripheral rural areas, decisions concerning the conversion of land and regional management should take account of their often complex and contrasting effects. They should be evaluated by using a procedural model that accounts for all the elements involved in the processes of regional transformation, both globally and according to their priority.

The evaluation methods which meet such requirements are the multi-criteria techniques. All these procedures rationalize the decision making process where choices are conditioned by a number of heterogeneous and contrasting objectives(Nijkamp and Voogd, 1989).

As already stated, it is essential that the first step of the evaluation process should determine some indicators expressing the various levels of value of the rural areas being examined.

In particular, the aspects concerning production may be expressed by means of an indicator including all the incomes generated by the production factors: this indicator is an *added value*, or it can be called *"regional macro-income" ("macroreddito territoriale")*(Grittani, 1988). *It consists of the sum of labour, capital and enterprise incomes, namely, of the overall wealth produced by the resource: "agricultural land".*

As far as the landscape values of the peripheral areas are concerned, it is important to clarify the meaning of "landscape", since different analytical and evaluation approaches can be chosen.

Landscape is the combination of various interacting environmental components. Its analysis and evaluation should be coherent and exhaustive with special regard to its defined meaning. When evaluating aspects of it, it is important to surmount the static concept that considers merely aesthetic values. As a consequence, the use of indicators emphasizing only these aspects is restrictive. However, it would be practical to analyze aesthetic aspects of the landscape by resorting to indicators which characterize some important features of the peripheral rural areas.

To this purpose some appropriate botanical indicators would be *the onset and the duration of flowering* and *the period for reconstituting plant species.* The first is obvious, as it refers to the importance of flowers in characterizing a landscape: they not only exalt the aesthetic qualities of a landscape but they are also one of the most impressive semeiotic features in the language of natural signs.

The indicator *period for reconstituting plant species* considers plant species as an environmental asset which takes decades or even centuries to grow and to reach a definite shape. Thus, their destruction is surely a considerable damage.

As an alternative another indicator could be used to express a monetary value: the *cost of reconstituting of destroyed species*, supposing that the area were to be restored to its previous conditions.

It has been said that peripheral rural areas can also be used as green areas for outdoor recreation. Some rural parks could be created to form a green belt around towns, in which where the peculiar rural features of the land as well as the traits of certain landscapes would be safeguarded and developed.

With regard to this, it would be very interesting to establish an indicator to quantifies the area of farmland to be preserved also in relation to the human pressure exerted on rural areas by urban spread.

A *location index* could be assumed to express the area of farmland to preserve within a given area (for example, in communes or bigger districts) and in relation to its resident population and its density.

A further indicator could express the accessibility of farmland on urban peripheries or within partly built-up areas.

The following paragraphs describe the city of Bari[1], in terms of a study that classified areas that had not been built-up and showed the varying degrees of conservation to be implemented. Its purpose was to show that, even in the most densely built-up areas, there are still some plots of farmland that could be preserved. This would satisfy the need to increase the number and extent of parks and gardens, as residents in Bari have no more than 1 m^2 of green space per capita.

The study has looked more at urban land than at rural land, even though, as already mentioned, some rural land was still used for farming. Its various locations have been identified and the following types of uses have been classified: market garden, combined garden, vineyard, olive grove, combined olive grove, and uncultivated land. Afterwards, the areas have been classified by means of the following indicators: 1) period in years to reconstitute plant species; 2) duration and type of flowers; 3) cost of restoring one square metre of the formerly green area; 4) location index expressed as functions of the area's physical and visual accessibility, of its area and distance from the central built-up area.

The degree of conservation to be implemented in the areas in question has been assessed by giving a score to each indicator class. Three degrees of conservation have been established (low, medium, high). The results of the study are shown in the map for the conservation of non-built-up areas (figure 1).

In particular, these unbuilt areas cover an area of approximately 1000 hectares: 15.5% accounts for vegetable crops, 48% for tree crops (mainly olives, *Olea europea*) sometimes grown together with other herbaceous species, and 36.6% uncultivated. Commercial cultivation of vegetable is located mostly along the coast south of Bari; trees, and in particular olive groves, are present in the hinterland to the northwest and southwest. The uncultivated areas are the sites reserved by the Master Plan (PEG Piano Regulator General) for urban spread.

The analysis of different uses of land and degrees of conservation (fig.1) shows that the areas with the highest degree of conservation are mostly olive groves. A high degree of conservation is considered also for most uncultivated areas, in relation to the importance of their location index.
An analysis of areas to be preserved shows the high degree has been set for 17.8% (189 ha), the medium degree for 42.5% (452 ha), and the low degree for 31.7% (337 ha).
According to the land uses envisaged in the Master Plan (PRG), 50% of these areas would be reserved for parks and gardens[2], slightly less than 10% for mining, agriculture and animal-farming, and the remaining 40% for industry and service sector (17,5%), for public facilities (15,5%), and for road-transport infrastructure (7%).
The results of the study indicate that there is little green space in the city of Bari, though the possibility exists that some cultivated and uncultivated areas may be adequate to meet the outdoor recreational needs of the public. It should also be noted that in this context the Public Administration would do only what is shown in the Master Plan (PRG). Indeed, half the examined area, 500 ha, is already reserved for urban parks and gardens. What should not be underestimated is the need carefully to plan these areas to develop both their existing natural resources and their remaining farming activities. Such considerations should be also extended to include all areas that have not been affected by urban transformation and which will not be used according to the Master Plan (PRG) in the near future. In the light of the new environmental requirements, the decisions and the choices of the plan should be reconsidered: Bari still has time to adopt new resolutions of problems of urban degradation and suburban areas in the light of a better understanding of the role that rural peripheral areas could play.

15.4 Conclusions

The "regional macro-income" ("macroreddito territoriale"), the period for reconstituting plant species, the type and duration of flowering, the cost of reconstituting plant species and the location indicators cannot fully express the value of peripheral rural areas, but they do offer some very pragmatic indications within a setting that ignores them completely.
Planners should make concrete and practical suggestions. Even a correct urban plan can fail if researchers do not co-operate with those who actually do the planning work. Gaps between theory and practice admit "eccentric" applications that are totally unrelated to theoretical principles. Theories in planning - as those in other practical disciplines -have to be tested to understand whether something is wrong with them.
For a long time in Italy, urban planners have spoken one language and professional operators another, mainly because no radical revision of the discipline had been made, even when environmental emergencies imposed a new planning approach.
The same is true of planners and evaluators. While new and more sophisticated evaluation methods are continuously proposed, the evaluators' value judgements have less and less bearing on theory.
Evaluators are not able to determine the public value of peripheral rural areas. They continue to evaluate them in private terms and assess favourably located farmland as higher and

peripheral land that is unbuilt as lower. Their scope of evaluation continues to be private, so a peripheral area can be either excellent farmland or a very unpleasant urban area.
Both city planners and evaluators see urban and rural areas as forming a continuum. An area can be either urban or rural: there is no intermediate condition. Thus the value of a peripheral rural area can exceed that of an urban area or fall short of that of a potential building site in the centre of the town.
As may easily be inferred, this ambiguity as to the identity of given areas affects the evaluation process. An urban area is always treated as a private area.
By contrast, in a broader planning framework, considering that the plan typically has a public connotation, evaluation should take account of public aspects. Planners and evaluators should consider the needs of not only some social groups but of the general public. Thus land in the countryside - albeit private property - must be considered a public resource that generates positive externalities that all can enjoy.
Thus a planning priority is to value a peripheral rural area as a public good, which poses a different question: the choice of the method to be used. Some evaluators argue that only monetary procedures are able to fully express the value of a public good; others believe that monetary techniques do not internalize its various functions. There is only a clear duality, no intermediate position.
In order not to show an generally-tolerant attitude, it is important to reject *a priori* positions. A multi-criteria procedure is undoubtedly more flexible and, thus, more appropriate when evaluating so complex a resource as peripheral farmland; but it should be noted that the value proposed by means of monetary methods is undoubtedly more immediately perceivable and, for this reason, can be easily accepted or refused.
Methods should be chosen according to context, starting from a basic assumption: whenever the evaluator can adopt monetary methods, he has the "moral duty" to do it. Now, this is true not so much because appraising science has always provided a monetary expression of value (traditions can be changed as well), but mainly because monetary methods are easy to understand, which is essential in evaluation.
Neither method can be accepted or refused *a priori*; the important thing is that the evaluator succeeds in carrying out the process of evaluation by means of objective hypotheses.

References

Brugnoli, A. (1990) "Valutazione dell'impatto urbano sulla struttura produttiva agricola", Genio Rurale 4, pp. 52-64

Grittani, G. (1988) "La valutazione monetaria del territorio rurale nei processi di pianificazione urbana e regionale", in A. Barbanente (ed.), Metodi di Valutazione nella Pianificazione Urbana e Territoriale. Teoria e Casi di Studio, Atti del colloqio internazionale, Capri-Napoli, CNR-IRIS, Bari

Lockeretz, W. (1989) "Secondary effects on Midwestern agricolture of metropolitan development and decreases in farmland", Land Economics, 65/ 3, pp. 205-216

Nijkamp, P. and Voogd, H. (1989) "Classificazione dei metodi di valutazione multidimensionali", in L. Fusco Girard (ed.) Conservazione e Sviluppo: la Valutazione nella Pianificazione Fisica, Franco Angeli, Milano, pp. 96-117

Toccolini, A. (1990) "Agricoltura periurbana e governo del territorio nel sistema metropolitano milanese", Genio Rurale 12, pp. 35-47

Zappavigna, P. and Tagliavini, I. (1990) "Politiche di piano e trasformazioni territoriali nella fascia periurbana parmense", Genio Rurale 4, pp. 65-73

Notes

[1] The study is included in Roberta Sisto's graduation thesis, "La valutazione delle aree di frangia: il caso di Bari", Facoltà di Agraria, anno accademico 1992-93, Università di Bari.
[2] The terms urban parks and gardens include open spaces in neighbourhood, gardens for the residents, and the parks and the gardens in the strict sense of the word, envisaged in the Master Plan (PRG).

Giovanna De Fano
Istituto di Estimo e Pianificazione Rurale
Facoltà di Agraria
Via Amendola 165/A
70126 Bari
Italy

16 A METHOD FOR THE EVALUATION OF A LARGE AREA: THE CASE OF CENTRAL APULIA SYSTEM

Sebastiano Carbonara

16.1 Conceptual framework of the analysis

Following the fundamental objective of this work, the research approach was based first on the analytical evaluation of the suitability and needs of the considered territory (figure 1), and a primary role was given to the knowledge of agricultural-environmental resources the conservation of which was considered to be the priority in defining appropriate land policies.

Preliminary theoretical-descriptive considerations about the possibility of defining a land-organization approach referred to discrete scopes recognizable on the basis of local peculiarities, led to identify, within the land reference area (the province)[1], specific scenarios to be confronted with in defining the intervention strategies to be calibrated on the intrinsic qualities of each of them.

This result enables us to go beyond the hierarchical-area-based interpretation of the territory where the mediation of the provincial capital (Bari) became a *sine qua non* for triggering virtuous dynamics. It has also focused attention on and given priority to site-specific and local situations, so as to avoid an object-view of the territory that depersonalizes the parties and transforms them into an undifferentiated whole with basically equal capabilities and prerogatives.

The need to depart from equalizing intervention approaches is also related to uncertain results of such a practice that, in the name of a rather undefined land re-equilibrium indistinctly pursued without considering the physical-social peculiarities of the concerned environments, has often directed past planning experiences.

Often governed by space and functional equalization principles, the objective of re-equilibrium was aimed in fact at creating conditions of general land indifference and at making the territory uniform in a way that would solve any imbalance by removing any form of diversity.

Rethinking such approaches necessarily involves interventions that take more care of the distinctive characters of the land in question.

With this in mind, rebalancing a land system defined by a set of single locations, of which each has its own structural characteristics that make it possible to define a differential land framework, means to set up a grid of interrelationships between the

D. Borri et al. (eds.), Evaluating Theory-Practice and Urban-Rural Interplay in Planning, 217–230.

different parties, through a better and more appropriate use of their historical, economic and cultural background.
The land differences will then become the factors on which a supra-municipal land planning should be based not to pursue static equilibrium objectives, but dynamic development processes".
In view of that, the theme of land re-equilibrium would lose its past ambiguity to move towards choices that take account of the conditions of each scenario previously defined on the basis of criteria that refer both to the area characteristics[2], and to privileged relationships between different, physically-separated subjects.

Figure 16.1 Representation of the 5 provinces of Apulia Region. The dark area is the province of Bari.

If such a condition reflects the need to use descriptive approaches capable of confronting with land organization forms based both on stable physical conditions and on relation (material and immaterial) flows, the present analysis has considered both gravitational and nodal hierarchical models and reticular approaches which are both present in describing complex land systems that integrate with each other to "give rise to an eclectic organization model of the urban framework"[3] organization.
The objective pursued has focused on the possibility of defining local systems that are characterized and recognizable as a result of these descriptions, from which one can infer operational assumptions that take account of the quality of each context, according to a strategic planning model that intentionally selects the projects in relation to the existing land situations.
Being a document of perspective, strategic organization and orientation interventions, this instrument also sets up the rules[4] and makes it possible to consider the territory as a diversified system.

In this perspective, it seems possible to define more efficient land re-organization policies as a result of a process based on the intra-institutional concerted action and cooperation, on the involvement of different social and economic actors, on the capacity to confront the market, in short, on a land integration that is not performed by neglecting diversities but rather by taking them as a basis for the strategies needed to trigger complementary and synergy.
Orientations on the state of supra-municipal planning in Italy, also after the law on the reform of Local Authorities[5] that has considerably enlarged the possibilities of interventions of the authorities intermediate between the Municipalities and the Regions, together with equalizing traditional, comprehensive approaches, based on urban-type prescriptions, seem to be oriented towards a strategic plan model for an urban development organized on multipolar networks and at exploitation of original structural characters (Pogliani, 1993) .
Since the strategic plan is flexible enough to be capable of integrating views of area and reticular organization, the attention to local specificities and the confrontation with relationships not necessarily constrained by physical contiguity criteria, it can be possible to propose the evaluative procedure as a form for legitimating the choices pursued, in that it would permit (according to different purposes that also refer to the objectives attainable through multi-criteria evaluative methodologies) an examination of issues relative to both the "choice of the best solution, which would follow a selection procedure, and the classification of actions chosen according to preferential orders, or a description of actions and/or their consequences according to the systematic and formalized modes or setting up a cognitive procedure"(Roy, 1985).
With this in mind, it would be possible to extend the meanings and the role of evaluation, by covering each stage of decision-making that defines the course of the plan. In particular, by giving priority to the decision-making steps and the *processus d'apprentissage* associated with them, the French multi-criteria methodologies would allow one, within the choices of the plan, to "clarify subjective aspects related to the individual systems of preferences, to the corresponding coherence levels, with the ultimate purpose of estimating the probability that a plan or a programme arises the consensus of a plurality of individuals (actors), each of them aiming at objectives which are often in conflict with each other"(Las Casas, 1992).
The possibilities offered for procedural innovations in the urban-planning instruments (the strategic plan) "would enable [it] to overcome the instrumental view of the role attributed in the planning activity to the evaluation practice which is no longer an internal component of the plan, separated from the identification of interventions and development guidelines in the urban area, but becomes a finalist component completely autonomous with respect to the traditional analysis"(Pucci 1992).
This paper reports the results obtained so far in the development of the empirical research carried out on the province of Bari and takes account of all the above remarks that aim to justify and motivate interest in the proposal of a evaluative model to be integrated with large scale plan policies.
Having aimed at:
- identifying areas homogeneous in terms of contextual specificity defined on the basis of criteria taken as fundamental to the purposes of the research;

it is now pursuing the aim of:

- checking the significance of a multi-criteria methodology within a evaluative process that develops first on the confrontation of the identified sub-areas - as a basic descriptive reference for the definition of orientations and strategies of territory governance - and then on the evaluation of the project options included within the planning instruments.

16.2 The analysis phase to date

The analysis has so far considered six monographs:
1) the study of the agricultural sector has followed an approach that has considered not only what is strictly speaking the production aspect but also the environmental character of rural areas;
2) the analysis of the economic system of the province;
3) the analysis of population and settlement evolution;
4) the eco-landscape analysis;
5) the hydro-geological analysis;
6) the analysis of the infrastructural transport system.

Once the existing reality had been decoded through specific keys, an attempt was made to reassemble the acquired sectorial knowledge within a global framework that could be sufficiently representative of the considered land system, first in descriptive-qualitative terms and then in synthetic-quantitative terms.

As previously said, the study-area covers the territory within the administrative boundaries of the province of Bari, and is the subject of a general institutional attention in the matter of environmental and land planning, considering that in this part of Apulia neither the boundaries of the *Citta' Metropolitana* (according to law 142/90)[6], nor the catchment basin of the Adriatic Coast have been defined yet, as stated by the Act 183/89 on the Defence of the Soil[7].

According to a classical schematization of the area made on the basis of the physical characteristics, a more or less sharp distinction can be made between the coastal strip and the medium-high hill strips of Murgia that run roughly parallel to the coast in a dominantly longitudinal north-west - south-east orientation (figure 2). The pedo-climatic factors are considerably more favourable on the coastal and sub-coastal strip where the flat or gently sloping fertile soil and the typical Mediterranean climate favour intensive agriculture that is extremely receptive to technological innovation. Going inland towards Murgia, this situation gradually changes as altitudes increase; winters are more severe and prolonged, and soils are poorer and exhibit frequent rocky outcrops. A constant factor is the almost total absence of surface water but the presence of notable ground-water resources related to the basically calcareus nature of the subsoil.

A purely descriptive approach to the geography of these places tends to emphasize a kind of land supremacy of rural environments. Agricultural activity still plays a major role in the province, not only in view of a significant economic output, but also of the notable surface area concerned which is equal to about 91% of the whole municipal territory.

In this case there is no marked erosion of the non-urban areas that largely maintain the prerogatives of structured agricultural schemes. Nevertheless, this should not be taken to imply that the Central Apulia System is free from the degradation of the countryside and natural spaces or that past planning choices had considered the environment as a

characterizing element. Indeed, effects have not been devastating as a consequence of an economic development model that concentrated demand of land for extra-agricultural purposes along the coastal strip, and only marginally involved the inland rural areas.
In particular, the increased conflicts in land use were evident within the territory of Bari and of the neighbouring municipalities, following a tendency that is not in decline. In the 1970s and 1980s, a notable reduction in the Agricultural Area (SAU = Superficie Agricola Utilizzata) was observed in almost all municipalities of the first and second ring, as well as in the capital of the province.
In addition to the presence of small-sized farms (scattered holdings are largely present especially along the coastal strip)[8], a strongly characterizing element of the agricultural production structure that considerably affects the landscape conformation of this territory, is also the massive presence of tree crops that occupy about 40% of the SAU. Olive trees are absolutely dominant (occupying slightly less than 2/3 of the whole arboreal area), although also table and wine vines and fruit orchards are markedly present.

Figure 16.2: Physical characterization of provincial territory according to pedo-climatic factors. The darkest shading shows the inland, higher-altitude Murgia area.

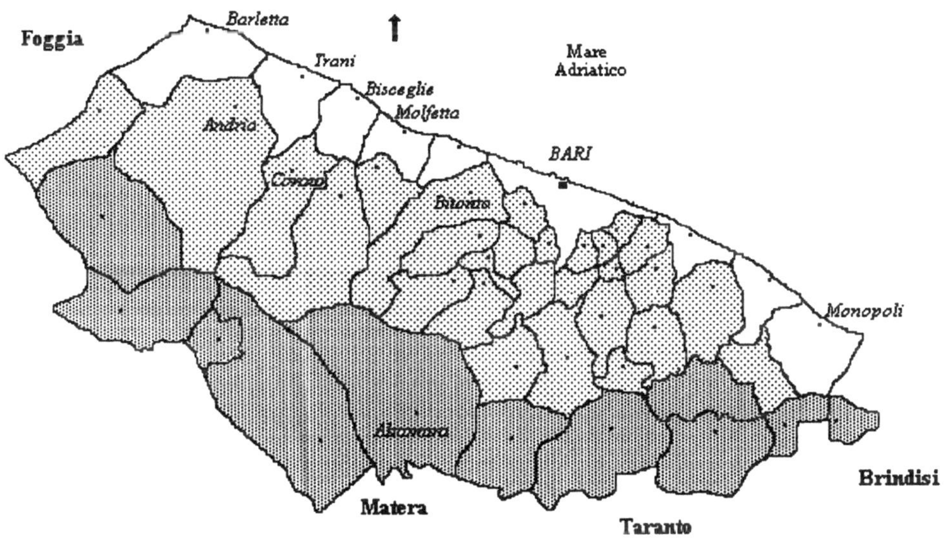

This extensive tree cultivation occupies mainly the agricultural surface between the coastal strip and the inland (stretching longitudinally in the province), up to m. 300-350 altitude, where cereals, rotational fodder, pastures and restricted forest areas almost completely replace any other form of cultivated species. The two areas could be respectively referred to as olive-grown and extensive cultivation and pasture areas". The latest trends in the agricultural sector emphasize generally steady markets for olives, vegetables and fruit production, and a consolidation of the animal-breeding activity south-east of Bari that is also integrated into a system of artisan- enterprises in the dairy-industry sector.

Although brief, the present information on the structural aspects should clarify the initial statements about the notable role of agricultural activity in the economy of the province; but one should also keep in mind its strength and its capacity to characterize the local geography and to mark the non-urban landscape in a way that still enables an observer to recognize tangible signs of peasant culture and civilization that still govern its transformation and play a vital role in preserving its hydro-geological equilibrium and maintaining the land spaces outside the built-up areas. However, it would be extremely misleading not to emphasize that the land weight of rural areas along the coastal strip has declined considerably since the 1960s.

As partly stated, the population growth of the province of Bari along the Adriatic coast has given rise to medium- and large-sized centres (mostly of industry and commerce) with access ensured by a well-structured transport network just along the coastal axis where the centrality of the capital of the province is evident; the region also contains a major port and airport.

Figure 16.3: Main road network along the internal areas of the province

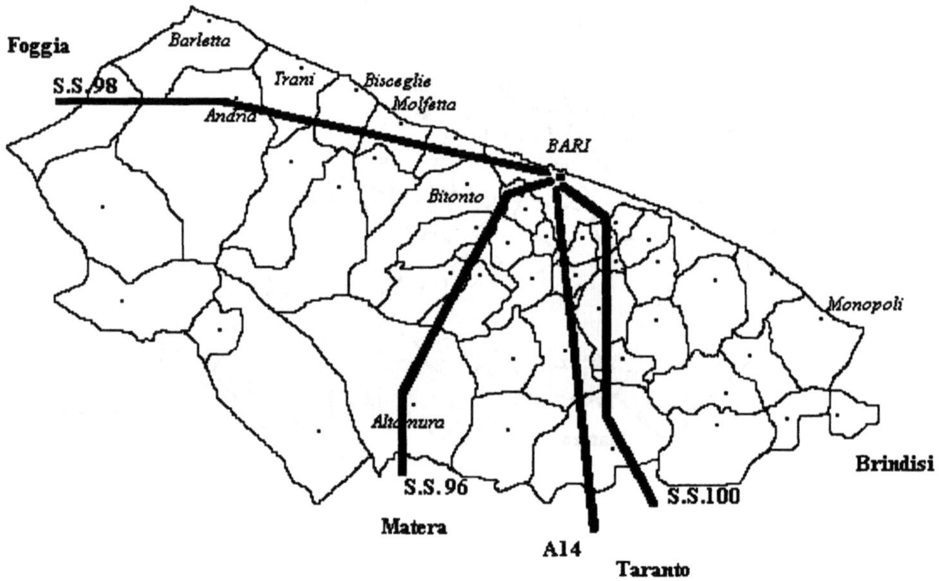

In fact, the trunk roads 16 and 16bis, the A14 highway and the State Railways lines Lecce-Ancona-Bologna, draw an intermodal corridor through the territory, that only marginally strengthens the interior of the province. In particular, the road network inland from the capital of the province follows three main axes (figure 3). These road sections, although underdimensioned with respect to their vehicular traffic, still remain the only transportation facilities capable of ensuring sufficiently rapid connections with Bari and along which the growth of new development centres is consolidating.

The imbalances one can observe in the road system are still more evident in the case of railways that, in terms of extension of the network and its performance, are quite insufficient to provide an adequate service along the transversal axes of the province.

This has allowed the coastal areas to achieve an economic development that is among the highest in Southern Italy, but which has jeopardized their environmental quality by creating conditions that could be properly defined as marginal by which we mean the areas characterized by irreversible land uses.

Although the degradation of coastal areas is often due to modifications inland, it is equally indisputable that what has occurred in the coastal vicinity of Bari has certainly not preserved its delicate and changeable natural balances: it has, instead, contributed to make its form and functions more rigid.

In most of this area, one may currently observe all the human factors that are considered to harm the environment: large urban settlements that have developed parallel to the coast, linear infrastructures near the coast, concentrations of polluting industrial activities, allowance of licences for tourist purposes without any preliminary analysis of environmental parameters, the abandonment and degradation of derelict industrial and rural facilities and so on.

The agricultural background is contained in an urban frame of 48 municipalities of which most are densely populated: in the hemispherical ring around Bari (population just over 340,000) 23 centres have populations over 20,000 and include towns with populations over 40,000 (86,000 for Barletta). Most towns (40) lie below 400 m above sea level.

Such an agglomeration comprises a discontinuous development of built-up areas where most urban fronts are clearly distinguishable from external areas, thus distinguishing Bari's metropolitan area from other Italian conurbations.

If the urbanization model in the province of Bari emphasizes the centrality of coastal municipalities, the existing socio-demographic and economic-productive problems seem to assign a less marginal role to the ancient rural inland centres. Therefore, in view of their development in the last few years, and of their relations with one another and Bari, a change in the original space-production relationships within the province may be perceived and the urban structure can be better represented, following the reticular criterion which implies going beyond a strictly hierarchical relational view.

In accordance with the prerogatives related to its status of chief town (that will become a metropolitan area in accordance with law 142/90), Bari actually keeps its leadership for the allocation of valuable functions although, in terms of demographic features, industrial activities and service provision, the land organization seems to be centred on four nodal areas, the central one being certainly the most important but not unique. Three additional agglomerations are spatially distinguishable as seats of urban functions: the Northern area (population about 220,000) containing Barletta-Andria-Trani; the Murgia area where Santeramo-Altamura-Gravina are the major centres (adjoining the Lucania region, 120,000), the South-Eastern area with Putignano-Noci-Alberobello (about 57,000), and Martina Franca, in the province of Taranto (figure 4).

In Bari, we have observed a progressive productive decentralization of the small industry (which has experienced an increase in the 1980s of 7,000 workers within the first ring municipalities) and of wholesale trade auxiliary-services firms. The process of functional reorganization has also concerned the commercial services of large-scale retail trade, which gave priority to the axes defined by internal lines, i.e. the trunk roads S.S. 100 towards Taranto (Casamassima and Gioia del Colle) and S.S. 98 towards Foggia (Modugno, Bitonto).

This could be supposed to have resulted in an explosion of the urban nucleus which has incorporated new districts more or less connected with the centre.

For some of them, a more complete process of integration has taken place or is still under way; although it fits into a clear strategy of metropolitan government, it is defined as an established trend.

Figure 16.4: Representation of the urban agglomerations of Bari territory

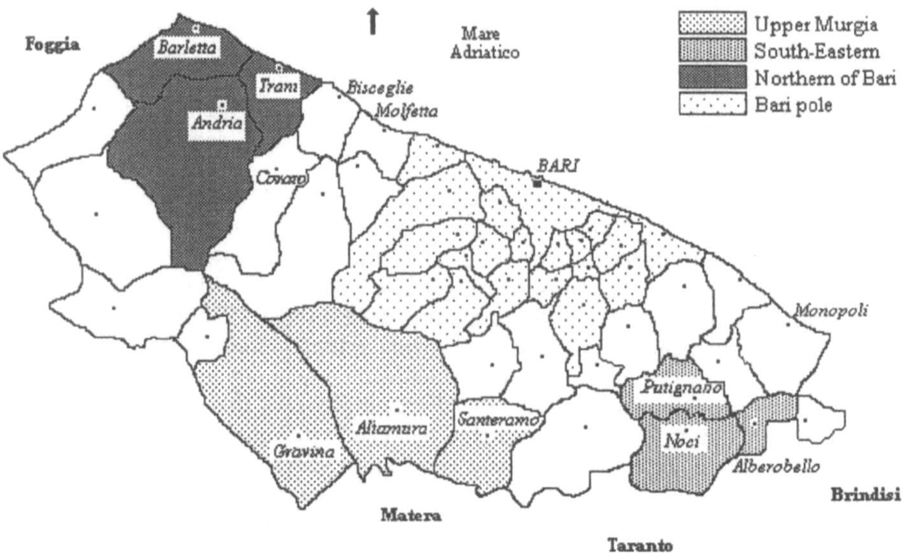

The observed settlement and economic evolution of the last decade seems to confirm the assumption of a current qualitative and quantitative change in the development model of the agglomeration of Bari province.

It stresses in particular the emerging role of the Murgia and South-Eastern areas where an increasing population has been observed that contrasts with conditions in the 1970s, together with a significant growth of manufacture activities so that these areas can be defined as the industrial poles of the province (as small and medium-sized enterprise specializing in wooden furniture and clothing) besides the consolidated poles of Bari and Barletta. However, what differentiates the two areas could be the service available to families and enterprises. In this connection, the South-Eastern area seems to be prevailing, as confirmed by the ratio of advanced-service workers/1,000 inhabitants[9], which is among the highest in the province. On the other hand, we should emphasize the

property development of the last decade in the Upper Murgia centres, that has contributed to the growth of the total housing stock of the province, to the same extent as the first and second ring municipalities.
It is indeed a marked percentage rather than an absolute growth on the aggregate datum, since in the 1980s a vast area of Bari experienced a decline that was in some way physiological, in property demand.
In spite of this, generally speaking, the shifting of settlements from coastal to inland areas seems to be progressively increasing and could be partly associated both to the markedly lower property values and to the existing environmental assets in inland areas.
In some respects, the economic development of the South-East and Murgia areas cannot easily be explained on the ground of the location criteria that are considered to be traditionally important for enterprises. They are indeed rather inaccessible territorial areas that are partly remote from the main road and railway lines of communication, lying mostly 300 m to 450 m above sea level in the upper and lower Murgia belts. This involves critically re-examining the thinking that applies the concept of marginal, in all its negative meanings, to mountain and hilly areas or, at least, questioning its unambiguous applicability.
Nevertheless and in spite of such relatively recent trends in economic development, the two mutually-related, productive, inland areas of Bari province still have a great vocation for agriculture and tourism.
On the grounds of the indications resulting from various investigations, some sub-areas were defined within the territory under study as being based on the four urban agglomerations in question. They were taken to be condensation nuclei of the remaining centres of the province, with respect to the territorial homogeneity levels defined in relation to different productive sectors, environmental specificities, according to the economic relations and the cultural and social similarities.

16.3 Identified sub-areas and indicators

The following areas (figure 5) were identified:
-Bari pole,
-North Bari,
-Upper Murgia,
-South-East.
The first step of the research concludes by defining, for each sub-area identified, fifteen indicators, expressed in the aggregate form. Their selection resulted from the analysis of the prerogatives assigned in the matter of land planning by the legislation to the Provincial Government. Our purpose was to devise indicators suited to defining land policies to be implemented through the appropriate planning tool, that is to say, the Coordination Land Plan, that which should indicate, in the words of the relevant legislation:
- "the different land uses in relation to the prevailing vocation of its parts;
- the preliminary localization of the major infrastructures and lines of communication;
- the guidelines for water, hydrogeologic interventions...
- the areas in which parks or natural reserves are recommended"[10].

Figure 16.5: Sub-areas defined within Bari territory.

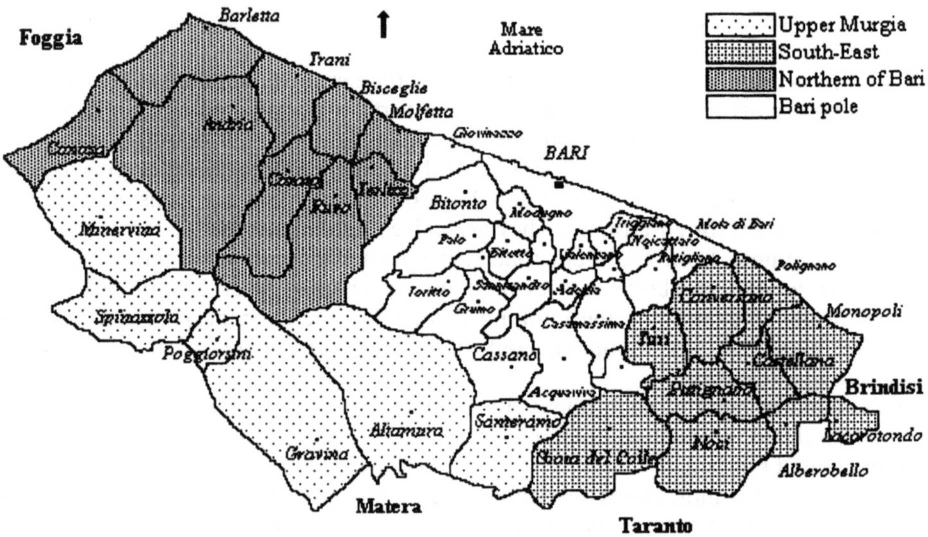

The indicators are:

-analysis of the agricultural sector:

1) the value added expresses the wealth produced by the agricultural sector in terms of gross saleable output excluding intermediate costs (technical inputs and services), referred to the unit area;

2) the degree of activity is the employment potential of the agricultural sector, expressed in terms of number of working hours/hectare;

3) the urban Agricultural Area: is a compound indicator based on classes of population density of urban areas that defines the share of the agricultural area (15 to 75 m^2 per capita as a function of density classes), which should be considered as applying to both rural space and public urban green assets;

-eco-landscape analysis:

4) the biopotentiality (Mcal/m2/a) is an eco-landscape indicator that refers to the latent capacity of homeostasis of an eco-system; it is derived from the notion of resistant stability referred to the main types of biosphere ecosystems and is based on the measure of metabolic or biomass data;

5) the grain-size measures the size of the spots that can define the micro-macro-heterogeneity in a landscape structure;

6. the index of species diversity: indicates the proportion in percent of the eco-mosaic in a landscape that is analysed by reference to the number of elements observed;

7) the stability identifies a range of classes that can be referred to the weight of landscape sub-systems of an eco-tissue, as influenced by their metastability threshold,

-hydrogeologic analysis:
8) index of environmental hazard, expressed in terms of disastrous climatic events that occurred in the area under study during the 1980s;
-analysis of the transport infrastructure system:
9) km of freeways/km2 of land area;
10) km of railway (excluding the narrow-gauge line/km2 of land area): both indicators are taken to express the degree of accessibility of each area and subsequently of their degree of resistance to transformation;
-analysis of the settlement system:
11) total estate resources;
12) unoccupied estate resources;
-analysis of the economic system:
13) per capita income indicator (in thousand liras): given by the ratio of the available income, calculated at current prices, to the resident population, that both expresses the population's economic activity and the welfare achieved;
14) non-agricultural enterprise density (local units/1,000 people), obtained from the aggregation of two indices, i.e. the industrial enterprise density (ratio of the number of local units in the energy, manufacture and building industrial sectors, to the resident population); such an indicator permits a dual interpretation by expressing the economic and entrepreneurial vitality of a given area and its entrepreneurial and productive potential by measuring its economic capability in a comparison with other territorial areas or with an area assumed as the target";
15) unemployment indicator (job seekers/100 resident people): this is a particular unemployment rate that can be understood as the global expression of social precariousness by quantifying the degree of difficulty of the working population in entering the job market; it takes account of the obsolescence of professional skills and the lack of a potential income.

16.4 The evaluative assumption

The multi-dimensional analysis of the issues concerned and the difficulty of grouping them into a single measuring expression suggested the choice of the evaluative model within the diversified family of multi-criteria analyses, just because the approach inherent in these analyses "does not imply the use of the monetary value to assess contradictory possibilities and is characterized by a greater flexibility under conditions which dictate concrete decisions"(Albers and Nijkamp, 1988).
By first approximation, we applied the concordance-discordance analysis, developed in France by B. Roy; it is a evaluative technique, set in the field of the decision theory. It is a hard multi-criteria analysis that permits a choice between a finite number of options and criteria, by using the values included in the evaluation matrix, adequately standardized in terms of efficiency, and the quantitative weights assigned to each comparison criterion.
It allows the alternative options to be arranged in accordance with different perspectives, and emphasized the one that best fits the criteria selected.
Preferences are measured through the binary comparison of alternative solutions that point out the advantage of the best alternative in relation to the one against with which it is

compared according to the number of criteria it responds to successfully. Such a comparison can basically express the relative superiority of one alternative to another, and is expressed by the concordance index.
In parallel to this, a discordance index, constructed through the binary comparison of alternatives, reveals the highest negative difference in relation to a criterion between the scores of two given alternatives that are to be compared; this thus indicates the soundness of the dominance relation achieved (the discordance is calculated based on the weighted and standardized matrix).
The second step consists in calculating indices of similarity between the concordance and discordance indices of each alternative and of two notional reference alternatives:
-a maximum alternative, constructed so as to satisfy all the criteria;
-a minimum alternative, based on the lowest scores for each criterion among different alternatives.
The relations between each concrete alternative and virtual alternatives can be established from the resulting classification.
If applied to our study, this briefly described evaluative procedure, would entail comparing the four territories in the area in question through the indicators of the six sectorial analyses and expressed in aggregate for each area.
Actually, the definition of the four sub-areas and the selection of the fifteen related indicators makes it possible to construct the evaluative matrix that can be interpreted as modelling the existing territorial reality.
In some ways, the application of the model to our specific case would imply a non-conventional use of it that would entail a comparison between alternatives (in relation to the possible scenarios defined based on criteria weighting) not to check preferable situations to support the firm choice of one of them but rather to provide a systematic and political priorities-oriented comparison.
In this step of the research, the evaluative procedure would involve a comparison of the areas concerned, based on the selected target criteria, weighed so as to give priority to environmental or economic, socio-demographic or infrastructure aspects, or to express indifference to each of them.
Such a comparison could represent, as a first step, the cognitive reference pattern supporting the setting of lines and strategic orientations of land governance; it could then become the evaluative support for qualitative and location choices of projects and interventions, and eventually include the implementation of the planning process.

References

Albers, L.H., Nijkamp, P. (1988) "Analisi multidimensionale per la valutazione dei piani e dei progetti. Come adeguare il giusto metodo al giusto problema" in: A. Barbanente (1988) International Symposium Proceedings Capri-Napoli, Metodi nella Pianificazione Urbana e Territoriale. Teoria e Casi di Studio

Las Casas, G.B. (1992) "Una ricerca di razionalita' a-priori nella valutazione dei piani", Territorio 12, Grafo Edition Brescia

Pogliani, L. (1993) "La pianificazione territoriale a scala provinciale. Metodologie e approcci a confronto", Territorio 14, Grafo Edition Brescia

Pucci, P. (1992) "Ruolo e funzione delle analisi valutative nella pianificazione urbana e territoriale", Genio Rurale 5, Edagricole, Bologna

Roy, B. (1985) "Méthodologie multicritère d'aide à la décision", Gestion, Paris

Notes

[1] In the Italian system, the Provinces represent the intermediate Bodies between the Regions and the Municipalities. The functions of these autonomous authorities (defined by general laws of the Republic) have recently been extended in the matter of land planning by the Act of June 8 1990, no. 142: *Ordinamento delle autonomie locali* (Regulations of local authorities). Among other things, this provision recognizes to the Provinces the faculty to prepare and adopt the *Piano Territoriale di Coordinamento* (Coordination Land Plan), a large-area planning instrument aiming at implementing environmental and land policies that cannot be pursued through the local scale plans alone.

[2] That is, in relation to natural factors (the physical size, the environmental, naturalistic, geomorphological components) or more specifically town-planning components, relative to settlement, production, residential background, to the organization of infrastructures and land use.

[3] Among the spatial organization approaches considered in the interpretation models of land phenomena, an interesting example is given by the proposals for the definition of *Citta' Metropolitana* (Metropolitan city), (ex lege 142/1990) which concern:
"- the approach to land homogeneity, distinguishable into urban-planning (or of the urbanized continuum) and structural-functional approach;
- hierarchical approach distinguishable into gravitational and nodal;
- the reticular approach". (Camagni R. Gibelli M.C., 1992, *L'area metropolitana milanese e la legge 142: un approccio economico-territoriale in termini di organizzazione a rete dei centri urbani*), in Territorio no. 11, page 130, Grafo edition Brescia.

[4] The reference is, in particular, to the French *Schéma Directeur* aiming at defining the wide strategic options, the definition of economic-land priorities, the check of the global consistency of the envisaged interventions, the definition of the rules within the medium - long term scenarios.

[5] See note 3.

[6] Articles 17ff., envisage the establishment of the *Citta' Metropolitana*, a new Body of intermediate size that specifically concerns only the biggest new Italian towns and neighbouring centres, of which they severally have "closely integrated relationships in terms of economic activities, services essential to social life, as well as cultural relationships and land characteristics". Defining boundaries of the metropolitan cities is entrusted to the Regions; as yet, they have not done this which, meanwhile, has changed from compulsory to facultative.

[7] Act of May 18 1989, no. 183 *Norme per il riassetto organizzativo e funzionale della difesa del suolo* (Rules for the organizational and functional readjustment of the detence of the soil). Article 13 subdivides the whole country into national, intraregional and regional catchment basins. In these areas the Catchment Basin Plan is in force. It has the value of "sectorial land plan and it is the cognitive, normative and technical-operational instrument through which actions and implementing rules are planned and programmed aiming at land conservation, defence and exploitation and at the correct use of waters, on the basis of the physical and environmental characteristics of the concerned territory".

[8] The mean SAU/farm ratio in the province is around 3.5 hectares.

[9] The advanced services considered for the purpose of the indicator include banking, finance and services provided to enterprises.
[10] Act n. 142/1990, art 151, subsection.

Sebastiano Carbonara
Istituto di Estimo e Pianificazione Rurale
Facoltà di Agraria
Università degli Studi di Bari
Via Amendola 165/A
70126 Bari
Italy

17 PLANNING IN URBANIZED AREAS UNDER NATURAL RISK

Francesco Gentile
Fabio Milillo
Giuliana Trisorio-Liuzzi

17.1 Preliminary remarks

The United Nations has sponsored research activity that focuses on reducing natural risks, mainly by targeting land planning on prevention activities, up to the year 2000.
All over the world, hundreds of towns and cities periodically suffer considerable damage and losses of human lives due to a series of natural phenomena of land instability. Being linked to specific, and sometimes synergetic, factors, they display their effects in different forms: landslides, seismic disturbances, avalanches, floods, volcanic eruptions, cyclones and other catastrophes.
Recurrence is a frequent and peculiar feature of these phenomena.
The magnitude of time of recurrence Tr (the average time span during which the event in question may be equalled or exceeded only one time), also referred to in literature as "return period", suggests that any planning process should carefully take account of all evaluations of the risk that such phenomena occur in the area being planned.
Although this might apparently be taken for granted, it is actually relevant for planning in the face of the current trend in public opinion and politics, to overlook such disastrous and damaging events and to treat them as exceptional.
All too many examples of destroyed or heavily damaged towns and cities and urbanized areas may be mentioned for each of these phenomena.
Italy has suffered from such natural disasters; from time to time, serious accidents have damaged even famous cities.
It would be impossible to list even the worst of them, but the few we mention will provide an exhaustive framework for the remarks that follow.
Ever since the late Middle Ages, the city of Florence has been threatened by the river Arno and has suffered many floods, which have influenced its urban planning: a compact shape surrounded by high walls.
Although the bridges — *Ponte Vecchio, Ponte alle Grazie* and *Ponte Santa Trinita* — were high and as reinforced as fortresses, the river kept on invading the city until the end

D. Borri et al. (eds.), Evaluating Theory-Practice and Urban-Rural Interplay in Planning, 231–245.

of the 18th century; it flooded streets and the ground floors of ancient palaces and destroyed bridges and defensive walls.

Table 17.1: Hazard of slow landslides.

Classes	and time of recurrence Tr (years)
I very high	**Tr < 2** continous seasonal lanslides
II high	**2 < Tr < 5** intermittent landslides
III mean	**5 < Tr < 20** intermittent landslides
IV low	**20 < Tr < 50** intermittent landslides
V very low	**Tr > 50** intermittent landslides

At the time, urban embankments were already built strongly in stone; they reduced the frequency of floods to one in one hundred years, but the risk of flooding persisted.
On 3 November 1844 a great flood inundated the city and the floodplain. On 4 November 1966 the citizens of Florence woke up in a flooded city. So many invaluable works of art were destroyed; the whole world was stunned by the event and many rushed to lend a hand in recovering the cultural heritage of the Renaissance (Ubertini, 1993).
Repeated floods in Calabria, Polesine and in Basilicata have occurred recently and, thanks to TV, the disasters suffered by Genova and other Italian cities have come to the eyes of the world.
Even Bari, the city hosting you, is subject to hydraulic risk (Puglisi et al., 1991). Until the twenties it was crossed by streams flowing to the sea, whose floods had caused inundations over centuries. The most dangerous is Picone; others include the Lamasinata, Montrone and Valenzano streams, which flow through the Adriatic side of Murgia (fig. 17.1).
Figure 17. 2 shows the areas of the city inundated by the disastrous floods in 1905, 1915 and 1926. That they were disastrous was a function of the town planning policy begun in the final decades of the 19th century.
The Picone stream had flooded in earlier centuries, but without the devastating consequences of the 20th century, because water invaded non urbanized areas. As for hydraulic risk, a significant town planning event was the reclamation of Mar Isabella (later contracted into "Marisabella"), a canal wanted by Duchess Isabella of Aragon to be dug in 1501 to change the town into an island. It was to surround the urban area, starting from the north-western coast - at the level of the present Corso Mazzini - and reaching the south-eastern coast. Construction started from the north-western side but was never

finished, yet it eased the flow of the Picone to the sea. Its reclamation led drive to a faster expansion of the town towards the Picone.

Figure 17. 1: Lamasinata, Picone, Montrone and Valenzano streams catchment areas and scheme of water management interventions carried out.

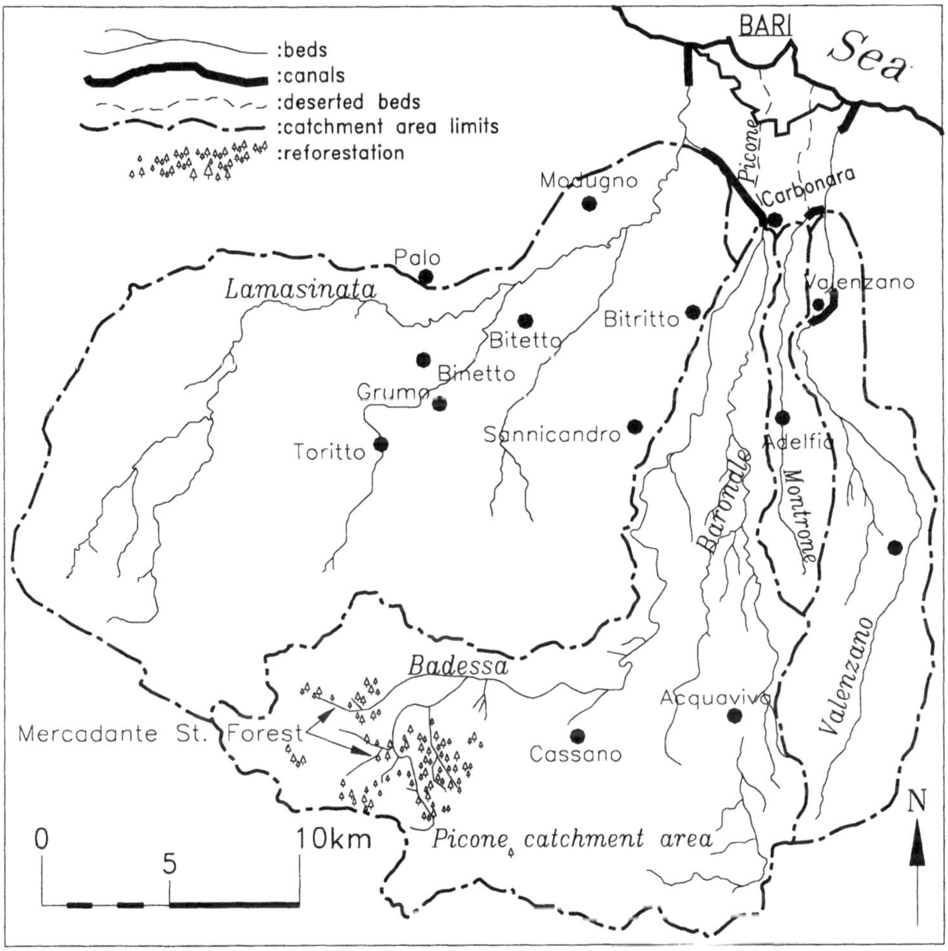

The local streams seemingly foster such careless choices. They remain dry for a long time; occasionally, they have a limited flow, so their beds are commonly and advantageously cultivated, for their soils have a higher water content than their surroundings. Proper floods are really rare and easily forgotten.

The first relevant urban flood of the Picone stream occurred in 1905. There were casualties, 150 houses were damaged, 4 thousand people were left homeless and lost all their belongings, the railway station was damaged, the Bari-Barletta tramline station was completely destroyed and one kilometre of its track was washed away. Damages were assessed at 3 million liras of the time.

Figure 17. 2:Map of the city of Bari reporting 1905, 1915 and 1926 flooded areas.

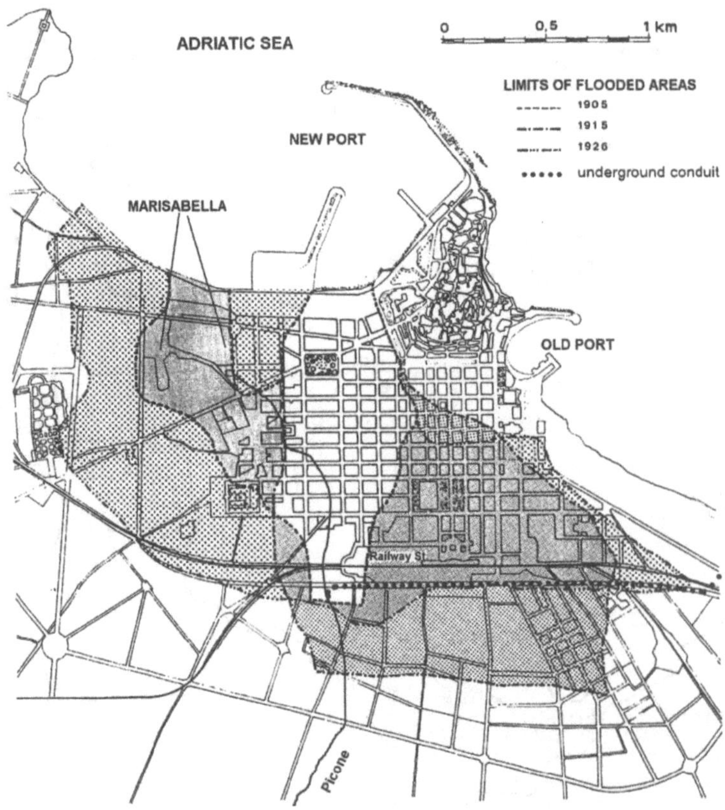

After the flood, the works to deviate the Picone into the Lamasinata stream were started: the project envisaged a weir on the Picone at Carbonara and a canal designed for a discharge of 160 m^3/s (Corpo Reale del Genio Civile, 1931).

The Picone and Valenzano streams flooded simultaneously in 1915, when the canal was not yet operational. Water heights in one hour ranged from 70 cm to 1 m. The increase of water heights in the town was caused in part by an existing sea quay that obstructed water outflow: explosive was used to breach it.

Further attempts were then made to protect the town from risk. The works started after 1905 were completed in the beginning of the twenties and an underground conduit was constructed to drain the small part of the watershed downstream of Carbonara (fig.2), which was thought to be the residual disturbance for the city. The old bed was filled and around it a new quarter rose.

The last and worst flood occurred in 1926, once again originated by the Picone. Its waters washed away the Carbonara weir, with an estimated discharge of 350 m^3/s, and resumed their previous bed; 20 people died and 65 were injured. Along the bed, between the Murgia and Bari, the flood crumbled some embankments.

Figure 17. 3: Isohyets concerning 5.11.1926 (after Alfieri, 1927) and 5.9.1957 events.

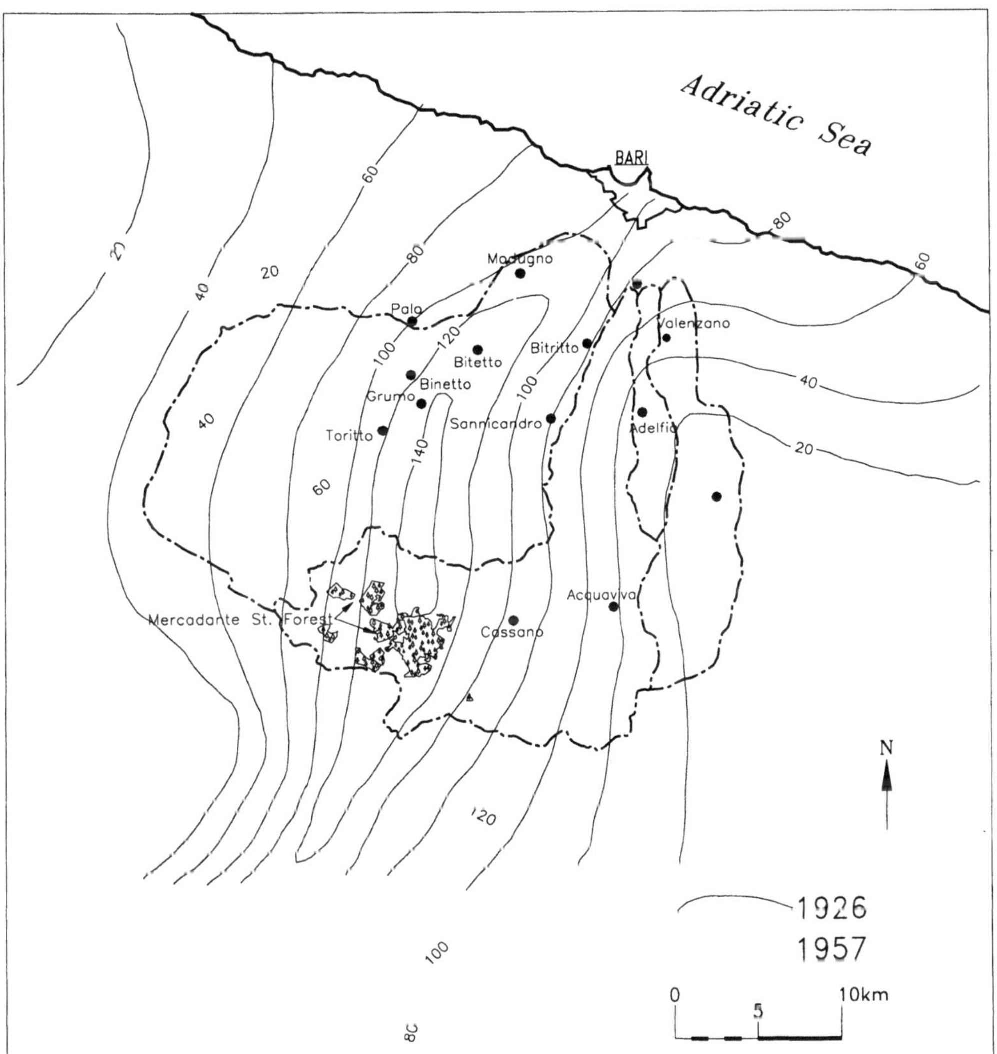

Figure 17.3 shows the isohyets of the rainfall preceding the 1926 flood and that on 5 September 1957; figure 17.4 gives the rainfall cumulated at two-hourly intervals before the floods in 1926 and 1957. These figures are useful in assessing the effects of the works done after the 1926 disaster.
The flood of 1957 in the Picone stream catchment area was perfectly controlled by the works built after 1926, although precipitation in the catchment area was higher than in 1926.
The management of watercourses running through the city of Bari started after the 1926 disaster and basically followed the scheme shown in figure 17.1.
This integrated watershed management scheme provides diversions, artificial channelling and weirs in the lower part of the catchment area, check dams at the head of the Picone, to prevent the flow of material into hydraulic works and gravelling of the bed, and finally reforestation (Mercadante State Forest), of a total surface in the Picone and Lamasinata catchment areas of about 1800 hectares.

Figure 17.4. Cumulative rainfall of 5.11.1926 and 5.9.1957.

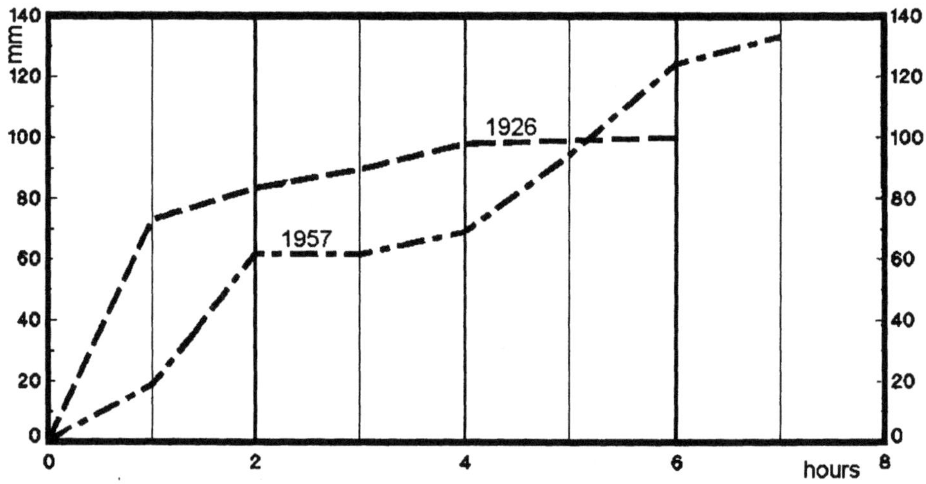

Today, these artificial canals, or *canaloni,* are cultivated, used as unauthorized dumping sites and cluttered with the embankments of access roads. The reasons why they were dug have been carelessly forgotten and the areas are now beginning to be referred to as locations to be somehow highlighted in a town planning perspective.
The situation in those areas, as described, calls for appropriate land use allocations that take due consideration of "risk levels".
Situations similar to that of Bari obtain in many other Italian coastal cities: Genova, Salerno, Reggio Calabria, Trapani, Manfredonia.
Besides these large scale phenomena, as specialists in the subject know well, buildings and urban areas, sometimes built in compliance with urban planning provisions, are

subject to flooding that, if not disastrous, certainly causes inconvenience and great economic losses.
Not to be underrated are some further consequences of water erosion forces and increasing damages directly related to floods: landslides, erosion.
Landslides are one of the most widespread and recurrent phenomena. Human lives and a considerable cultural, environmental and architectural heritage is often jeopardized; it is enough to recall the notorious cases of the Valley of the Temples at Agrigento, and Assisi, Orvieto, Ancona and many others in the whole Italian peninsula and its islands.
Landslides are of different types and modes, vary in duration and dimension, and follow different evolutionary paths as a function of a great number of variables, ranging from the nature of the formation to its related tectonic history.
Given this wide variety, we refer below to landslides that are subject to slow kinematics and intermittent reactivation.
Hundreds of towns in the Apennines, and 116 of 131 in Basilicata (Southern Italy) alone, are threatened by landslides of this type (Del Prete et al., 1992).
The instability of these areas, with a high concentration of landslide masses inherited from previous unfavourable periods, has been worsened by the phenomena of building expansion and migration to these towns over the last century, adding disturbing human activities to natural factors.
Natural slopes that have survived undamaged for thousands of years of various intensity landslide phases very rarely give rise today to original movements. On the contrary, generally weak areas, already hit by landslides in the past, periodically move as a consequence of intense meteoric phenomena, seismic shaking or careless human activities. The periodical repetition of landslide phenomena at the same site when major meteoric events hit these areas has been widely reported over the last century, and when the most recent earthquakes occurred. During the 1980 earthquake in Campania and Basilicata, excepting a few collapses, the majority of landslide movements turned out to be pre-existent.
Campomaggiore, in the province of Potenza, in the Basilicata region, is a revealing tale. During the year 1888, the ancient town was deserted as the landslide plate it had been built on had started to move again.
The list might go on mentioning a series of more or less famous towns and cities threatened by seismic events, volcanoes, avalanches and so on.
Consequently, great attention should always be devoted to land and water protection problems, from identifying them, planning measures of stabilization and control and organizing civil defence systems (structural and non structural interventions).

17.2 Criteria for the inclusion in planning schemes of areas subject to risk generated by natural phenomena

Up to now the problem has been examined mainly from the point of view of:
-soil and water protection in areas subject to risk (stabilization, conservation, control by structural interventions);
-soil destination and use obligations (the Italian legislation is complying to this approach with targeted norms aiming at management and protection of the land and its resources);

-rebuilding on new sites when stabilization costs are too high;
-planning warning and civil defence systems.

As for stabilization and protection of areas subject to risks generated by natural phenomena (structural defence measures), the productive use or re-use of these areas however depends on restoring acceptable conditions of stability, that are ensured by admissible safety coefficients obtained by means of appropriate measures.

Planning has often considered such risk areas to the extent of forbidding their occupation.

Planning non structural warning and civil defence systems derives from the need to integrate structural and non structural urban protection measures, at least until structural projects can reduce risks to insignificance.

Recurring to the example provided by Florence, hydrological studies and recent experience have well defined the river Arno.

A Pilot Project for the river aims at increasing water retention in its catchment area and to limit the risk for high water. Some scientific studies have been made about the genesis and propagation of floods in this area and its outflow capacity has increased from 2900 to 3200 m^3/s (but still lower than the maximum value of 4300 m^3/s estimated for the 1966 flood) as the thresholds of the *Ponte Vecchio* and *Ponte Santa Trinita* bridges have been lowered of one metre.

On the basis of the available data, it has been calculated that the recurrence time of a flood of the extent of that of 1966 is 120 years. The lowering of bridge thresholds might increase it to 150.

The works proposed by the Pilot Project have only recently been started and have so far concerned the construction of the Bilancino dam in the upper tract of the river Sieve.

It has therefore become necessary to integrate the structural works with the Civil Defence Plan (Ministry of Civil Defence; Toscana Region; Municipality of Florence and other local authorities; CNR, National Research Council, - GNDCI, National Group for Hydrogeological Disasters Protection), which envisages a flood-warning system with a presumptive prediction of three hours (a time span that several studies in progress have successfully increased) and an intervention plan for utility services, civil servants, police forces and citizens (Ubertini, 1993).

Two points of importance to the planning process emerge from the foregoing.

The first is coordinating operational decisions and, thus, the actual cooperation between planners, State and scientific bodies dealing with non structural defence measures, with specific reference to arranging escape and evacuation routes. Unfortunately, very large areas are somehow doomed to disaster as building and town planning activities in recent decades have made it very difficult to device appropriate evacuation plans (besides several areas subject to hydraulic risks, consider only the areas at risk from eruptions of Vesuvius or Etna).

The second point is the degree of safety obtained in any landscape that, although subject to risk generated by natural phenomena, has been targeted for leisure or productive activities.

Such activities must be compatible with the related risks, which should be determined and classified as a function of the features of each relevant phenomenon in terms of evolution over time, current probable activity induced by natural and/or human factors and possible damage caused by the event itself.

In any given urbanized area or suburb, identifying zones of different risk levels or degrees can prompt further measures to prevent or limit direct or indirect damage.
This framework can include agriculture, pasture and forestation, appropriately classified by type and organizational structure as functions of safety in relation to risk levels and to the environmental situation, and can cause areas otherwise doomed to disuse to become productive in terms of revenue or as attractive landscapes.
This approach may therefore be appropriate as a means of protecting and preserving the soil economically, ecologically and aesthetically.
Only apparently simple, this causes a number of problems about including the evaluation of risk areas in the planning process.
In Italy, some maps of landslide areas have been drawn up in recent years, beginning in the 1970s, mainly with the Targeted Project "Soil conservation" and with the CNR (National Research Council) "National Group for Hydrogeological Disasters Protection" (GNDCI).
A recent UNESCO report (Varnes et al., 1984) states the need to unify at international level the meaning of the terms hazard, vulnerability and risk.
H, natural hazard, indicates the probability that a given, potentially detrimental phenomenon might occur in a given area within a given time span.
V, vulnerability, is the degree of loss of a given element or set of elements resulting from the occurrence of a natural phenomenon of a given magnitude. It is therefore expressed in a range varying from 0 (no damage) to 1 (total damage).
Rs, specific risk, is the expected degree of loss due to a specific natural phenomenon:

$$Rs = H \times V$$

The elements E exposed to the risk are the total value of population, properties and economic activities, including public services, subject to risk in a given area.
Total risk R is the expected number of fatalities, injured persons and damage to property or destruction of economic activities due to a specific phenomenon and this is associated to an hazard of a given magnitude.
It is therefore the product of specific risk, Rs, times the elements exposed to the risk, E:

$$R = E \times Rs = E\,(H \times V)$$

The ZERMOS project in France has been improved, since 1984, by the State financed drawing of PER (Plan d'exposition aux Risques) mapping, concerning landslides, avalanches, floods etc.
In the framework of the different phases of the project, regional preliminary hazard zones (scale 1:100.000) and maps of evaluation of local hazard CR (1:10.000 - 1:5.000) have been followed by analyses of vulnerability based on an economic assessment of the area and existing works. The obtained PER have legal force for planning and building construction (Del Prete et al., 1992).
An example of application of this method may be that of towns or cities located in mountain or hilly areas that are prone to slow kinematics landslides, whose risk of occurrence can relate directly to the frequency of the reactivating causes.

The problem of quantifying such risks has been overcome by devising a method based on the evaluation of hazard and vulnerability elements, after having analyzed separately all geometrical features, kinematic elements and reactivating causes of typical landslide movements.
This method has been applied to two typical landslide areas: the area of Ferrandina (Lucanian foretrough) in the province of Matera (Basilicata region) and the area of Avigliano (Apennines chain) in the province of Potenza (figure 17.5) (Del Prete et al., 1992).

Figure 17.5: Location of Ferrandina and Avigliano

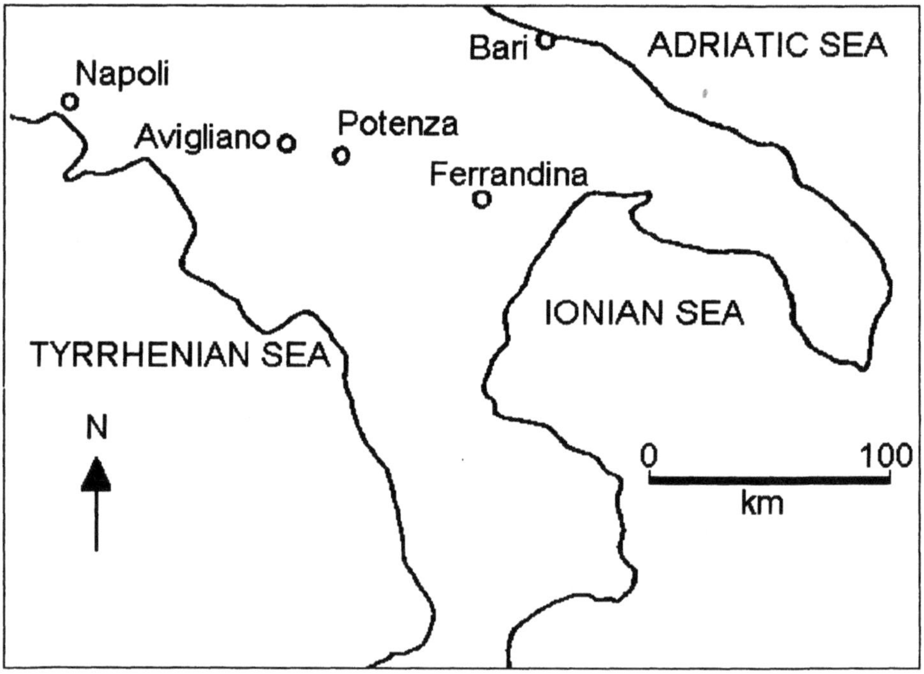

As for evaluating the hazardousness of landslide bodies existing in the two villages, that is the probability that they will reactivate within a given time, activating factors have been synthetically related to seismic and meteoric events and to human interferences:

$$Ht = Hs + Hm + Ha$$

where Hs is hazard related to seismic events, Hm hazard related to meteoric events and Ha related to human action.

The element that provides evidence for the overall hazard of a landslide within a given time derives from the evaluation of the present mobility state and of the periodical nature of reactivation causes over time by means of direct morphologic analysis and collection of historical data.

Thus, in the table 1 (Del Prete et al., 1992):

class I includes continuous and seasonal landslides with recurrence times of two years or less and indicates all currently unstable forms that may become active again without specific disturbance;

class II, high hazard, includes intermittent landslides with recurrence times between two and five years. Some might even fall within class I, as reactivating events may not be considered as exceptional, yet they recur less constantly than class I events and, for instance, seasonal rainfall variations alone are not enough to set them off;

class III, medium hazard, includes landslides having a reactivation intermittence ranging from five to twenty years. The relevant reactivation phenomena are exceptional, being linked to pluviometric or seismic events.

The same applies to the following classes, for which activation causes are even more exceptional, both in relation with heavy rainfall and seismic events.

On the basis of prediction analysis by deterministic or statistical methods, hazard maps have been drawn, locating landslides on the I-to-III scale.

As for related hazard, it seems appropriate to make a brief reference to hazard caused by human activities.

Disturbances to tension states of landslide bodies caused by human interventions are particularly detrimental in highly urbanized areas or areas undergoing an urbanization process.

Some of the most detrimental and yet widespread practices are: stripping and/or loading, often carried out in crucial areas of landslide bodies, alterations in underground hydraulic regimes deriving from the inflow of water into the landslide (via waterworks and sewage systems), removal of vegetation and vibrations due to explosions.

Assessing increases of hazard due to human activities that influence landslide bodies should be done in terms of impact of works on landslide areas.

This cannot be considered as cyclic (like rainfall or seismic events) but calls for analyses of local situations in terms of town planning.

A telling case is that of another village in Basilicata, Senise, where a landslide suddenly occurred on 26 July 1986 at 4 a.m. It killed 8 persons and destroyed 10 houses (Del Prete and Hutchinson, 1988).

According to the Authors, its main cause was wide stripping along the toe area to build three adjacent blocks of flats. It is a typical case of stripping in an urban expansion area along an actually or potentially unstable slope (figure 17.6).

Vulnerability, or the amount of damage determined by landslide activity, was calculated by a formula that took account of many factors, including numbers of buildings and inhabitants and costs of structures and infrastructure.

Such formula may certainly be modified to fit local situations in order to express at best the overall value of the goods at risk.

A sequence of maps were drawn for the two villages mentioned: besides maps of urban areas subject to landslide movements, also maps for hazard, vulnerability and therefore risk.

Figure 17.6: Senise: plan view indicating the position of the breaking line at the toe on the occasion of the mass movement of July, 26, 1986 (after Del Prete and Hutchinson, 1988).

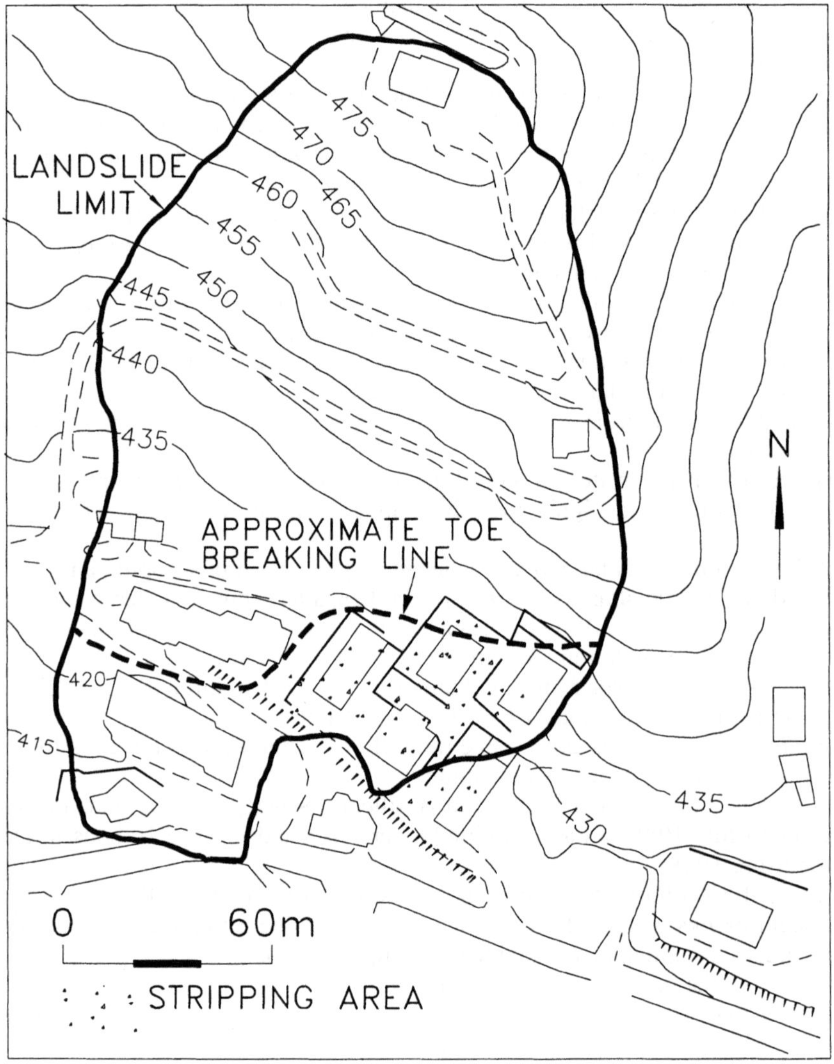

Figure 17.7: Diagram of the classes of hazard (H), vulnerability (V) and landslide risk (R) for the town of Ferrandina (after Del Prete et al., 1992).

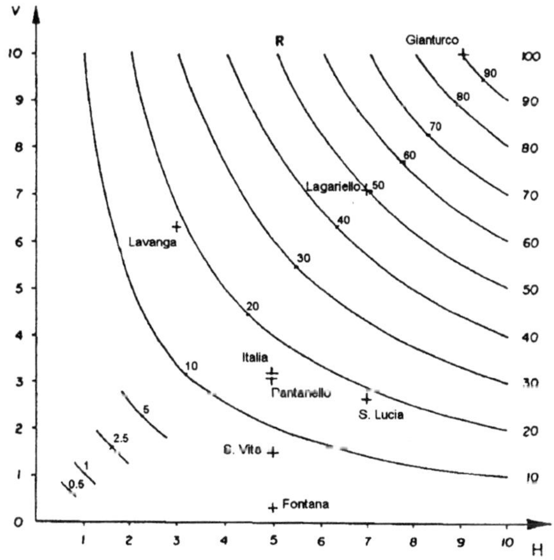

Figure 17.8. Diagram of the classes of hazard (H), vulnerability (V) and landslide risk (R) for the town of Avigliano (after Del Prete et al., 1992)

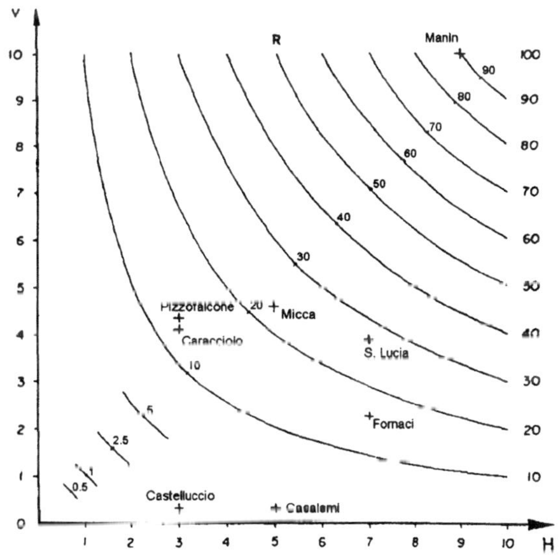

Risk indexes referred to each landslide in the urban area of Ferrandina and Avigliano are shown in figures 17.7 and 17.8, as calculated. On the basis of these calculations, they show in a simplified form curves indicating locations of equal risk, having assumed, on the basis of a given procedure, a dimensionless risk index ranging from 0 to 100.
A landslide area may thus be analyzed on the diagram for vulnerability and hazard, even with equal values for global risk.

17.3 Conclusions

In recent years, a number of thematic maps have been drawn to represent the condition of urban and rural areas subject to risk generated by natural phenomena.
Precious as this information may be, the location of areas subject to hazardous natural phenomena is still insufficient.
We therefore propose more attention during planning processes to identify zones with different degrees of risk (calculated-risk zones); after that, by forecasting possible direct or indirect damages, such zones could be put to new use.
Thus, agriculture, pasture and forestation could fit into those "calculated-risk zones" within urban or rural areas, that would otherwise be doomed to disuse, turning them into productive zones in terms of revenue and landscape.
Such activities may therefore be appropriate means of economical, aesthetical and ecological land protection; urban and rural planning in areas subject to risk generated by natural phenomena might take account of this opportunity.

References

Alfieri, S. (1927) L'alluvione della zona barese nel novembre 1926. Annali del Ministero dei Lavori Pubblici, Bari

Corpo Reale del Genio Civile (Royal Engineers Corps) (1931) "Le Opere del Regime per difesa della città di Bari dalle inondazioni", Laterza e Polo, Bari

Del Prete, M., Giaccari, E. and Trisorio-Liuzzi, G. (1992) "Rischio da frane intermittenti a cinematica lenta nelle aree montuose e collinari urbanizzate della Basilicata", Rapporto CNR-GNDCI, U.O. 2.37, Potenza

Del Prete, M. and Hutchinson, J.N. (1988) "La frana di Senise del 26.7.1986 nel quadro morfologico del versante meridionale della collina Timpone", Rivista Italiana di Geotecnica, anno XXII, n.1

Puglisi, S., Arciuli, E. and Milillo, F. (1991) "Il ruolo primario delle sistemazioni idraulico-forestali nella difesa di Bari dalle inondazioni", Monti e Boschi, n.1

Ubertini, L. (1993) "Gestione dei Rischi Idrogeologici in Italia Centrale: il "Progetto Arno", STOP Disasters, n.13, Osservatorio Vesuviano, Napoli

Varnes, D.J., et al. (1984) Landslide hazard zonation: a review of principles and practice. UNESCO pub., Paris, n. 63

Francesco Gentile
Fabio Milillo
Giuliana Trisorio-Liuzzi
Istituto di Sistemazioni Idraulico-Forestali, Facoltà di Agraria
via Amendola 165/a
70126 Bari
Italy.

The GeoJournal Library

1. B. Currey and G. Hugo (eds.): *Famine as Geographical Phenomenon.* 1984 ISBN 90-277-1762-1
2. S.H.U. Bowie, F.R.S. and I. Thornton (eds.): *Environmental Geochemistry and Health.* Report of the Royal Society's British National Committee for Problems of the Environment. 1985 ISBN 90-277-1879-2
3. L.A. Kosiński and K.M. Elahi (eds.): *Population Redistribution and Development in South Asia.* 1985 ISBN 90-277-1938-1
4. Y. Gradus (ed.): *Desert Development.* Man and Technology in Sparselands. 1985 ISBN 90-277-2043-6
5. F.J. Calzonetti and B.D. Solomon (eds.): *Geographical Dimensions of Energy.* 1985 ISBN 90-277-2061-4
6. J. Lundqvist, U. Lohm and M. Falkenmark (eds.): *Strategies for River Basin Management.* Environmental Integration of Land and Water in River Basin. 1985 ISBN 90-277-2111-4
7. A. Rogers and F.J. Willekens (eds.): *Migration and Settlement.* A Multiregional Comparative Study. 1986 ISBN 90-277-2119-X
8. R. Laulajainen: *Spatial Strategies in Retailing.* 1987 ISBN 90-277-2595-0
9. T.H. Lee, H.R. Linden, D.A. Dreyfus and T. Vasko (eds.): *The Methane Age.* 1988 ISBN 90-277-2745-7
10. H.J. Walker (ed.): *Artificial Structures and Shorelines.* 1988 ISBN 90-277-2746-5
11. A. Kellerman: *Time, Space, and Society.* Geographical Societal Perspectives. 1989 ISBN 0-7923-0123-4
12. P. Fabbri (ed.): *Recreational Uses of Coastal Areas.* A Research Project of the Commission on the Coastal Environment, International Geographical Union. 1990 ISBN 0-7923-0279-6
13. L.M. Brush, M.G. Wolman and Huang Bing-Wei (eds.): *Taming the Yellow River: Silt and Floods.* Proceedings of a Bilateral Seminar on Problems in the Lower Reaches of the Yellow River, China. 1989 ISBN 0-7923-0416-0
14. J. Stillwell and H.J. Scholten (eds.): *Contemporary Research in Population Geography.* A Comparison of the United Kingdom and the Netherlands. 1990 ISBN 0-7923-0431-4
15. M.S. Kenzer (ed.): *Applied Geography.* Issues, Questions, and Concerns. 1989 ISBN 0-7923-0438-1
16. D. Nir: *Region as a Socio-environmental System.* An Introduction to a Systemic Regional Geography. 1990 ISBN 0-7923-0516-7
17. H.J. Scholten and J.C.H. Stillwell (eds.): *Geographical Information Systems for Urban and Regional Planning.* 1990 ISBN 0-7923-0793-3
18. F.M. Brouwer, A.J. Thomas and M.J. Chadwick (eds.): *Land Use Changes in Europe.* Processes of Change, Environmental Transformations and Future Patterns. 1991 ISBN 0-7923-1099-3
19. C.J. Campbell: *The Golden Century of Oil 1950–2050.* The Depletion of a Resource. 1991 ISBN 0-7923-1442-5
20. F.M. Dieleman and S. Musterd (eds.): *The Randstad: A Research and Policy Laboratory.* 1992 ISBN 0-7923-1649-5
21. V.I. Ilyichev and V.V. Anikiev (eds.): *Oceanic and Anthropogenic Controls of Life in the Pacific Ocean.* 1992 ISBN 0-7923-1854-4

The GeoJournal Library

22. A.K. Dutt and F.J. Costa (eds.): *Perspectives on Planning and Urban Development in Belgium.* 1992 ISBN 0-7923-1885-4
23. J. Portugali: *Implicate Relations.* Society and Space in the Israeli-Palestinian Conflict. 1993 ISBN 0-7923-1886-2
24. M.J.C. de Lepper, H.J. Scholten and R.M. Stern (eds.): *The Added Value of Geographical Information Systems in Public and Environmental Health.* 1995 ISBN 0-7923-1887-0
25. J.P. Dorian, P.A. Minakir and V.T. Borisovich (eds.): *CIS Energy and Minerals Development.* Prospects, Problems and Opportunities for International Cooperation. 1993 ISBN 0-7923-2323-8
26. P.P. Wong (ed.): *Tourism vs Environment: The Case for Coastal Areas.* 1993 ISBN 0-7923-2404-8
27. G.B. Benko and U. Strohmayer (eds.): *Geography, History and Social Sciences.* 1995 ISBN 0-7923-2543-5
28. A. Faludi and A. der Valk: *Rule and Order. Dutch Planning Doctrine in the Twentieth Century.* 1994 ISBN 0-7923-2619-9
29. B.C. Hewitson and R.G. Crane (eds.): *Neural Nets: Applications in Geography.* 1994 ISBN 0-7923-2746-2
30. A.K. Dutt, F.J. Costa, S. Aggarwal and A.G. Noble (eds.): *The Asian City: Processes of Development, Characteristics and Planning.* 1994 ISBN 0-7923-3135-4
31. R. Laulajainen and H.A. Stafford: *Corporate Geography.* Business Location Principles and Cases. 1995 ISBN 0-7923-3326-8
32. J. Portugali (ed.): *The Construction of Cognitive Maps.* 1996 ISBN 0-7923-3949-5
33. E. Biagini: *Northern Ireland and Beyond.* Social and Geographical Issues. 1996 ISBN 0-7923-4046-9
34. A.K. Dutt (ed.): *Southeast Asia: A Ten Nation Region.* 1996 ISBN 0-7923-4171-6
35. J. Settele, C. Margules, P. Poschlod and K. Henle (eds.): *Species Survival in Fragmented Landscapes.* 1996 ISBN 0-7923-4239-9
36. M. Yoshino, M. Domrös, A. Douguédroit, J. Paszynski and L.D. Nkemdirim (eds.): *Climates and Societies – A Climatological Perspective.* A Contribution on Global Change and Related Problems Prepared by the Commission on Climatology of the International Geographical Union. 1997 ISBN 0-7923-4324-7
37. D. Borri, A. Khakee and C. Lacirignola (eds.): *Evaluating Theory-Practice and Urban-Rural Interplay in Planning.* 1997 ISBN 0-7923-4326-3

KLUWER ACADEMIC PUBLISHERS – DORDRECHT / BOSTON / LONDON

MIX
Papier aus verantwortungsvollen Quellen
Paper from responsible sources
FSC® C105338

If you have any concerns about our products,
you can contact us on
ProductSafety@springernature.com

In case Publisher is established outside the EU,
the EU authorized representative is:
Springer Nature Customer Service Center GmbH
Europaplatz 3, 69115 Heidelberg, Germany

Printed by Libri Plureos GmbH
in Hamburg, Germany